ILLUSTRATED HUMAN EMBRYOLOGY

VOLUME I

EMBRYOGENESIS

ILLUSTRATED HUMAN EMBRYOLOGY

VOLUME I: **EMBRYOGENESIS.**

VOLUME II: **ORGANOGENESIS.**

VOLUME III: **NERVOUS SYSTEM AND ENDOCRINE GLANDS.**

ILLUSTRATED HUMAN EMBRYOLOGY

VOLUME I

EMBRYOGENESIS

by

H. TUCHMANN-DUPLESSIS, M.D., Ph.D.
Professor, University of Paris Medical School
Paris, France

G. DAVID, M.D.
Associate Professor, University of Paris Medical School
Paris, France

P. HAEGEL, M.D.
Assistant Professor, University of Paris Medical School
Paris, France

TRANSLATED BY

LUCILLE S. HURLEY, Ph.D.
Professor, University of California, Davis, California

SPRINGER VERLAG
NEW YORK

CHAPMAN & HALL
LONDON

MASSON & Cie, ÉDITEURS
PARIS

1975

ISBN *0.387.900.18.7*
3.540.900.18.7
0.412.107.00.7

 Library of Congress Catalog Card number 72 177 236

INTRODUCTION

EMBRYOLOGY *studies the succession of transformations undergone by the fertilized egg in the formation of a new individual. Development of the embryo is directed by morphogenetic mechanisms ruled by a strict chronology. Survival of the egg, its transport in the genital tract, and the adaptation of the maternal organism to its presence are controlled by hormonal actions.*

Knowledge of these subjects is proving to be increasingly important for the medical practitioner. Such information helps to explain anatomic correlations; organ relationships also illuminate the etiology of numerous pathologic conditions. Disturbances of prenatal development engender congenital malformations and constitute an important cause of perinatal mortality and postnatal morbidity.

Our goal in preparing Volume I was to introduce the student to the complex phenomena of embryonic development in a clear and direct way.

We tried to present an overall view of the simultaneous nature of the multiple and rapid events in embryogenesis. Also, we thought it appropriate to treat certain topics in depth, because of their difficulty, or their physiological or pathological implications.

Thus, gastrulation and formation of the body shape, including flexion, which are particularly dynamic processes, can best be understood by following their sequences in detail. Special attention was also devoted to the placenta, since it assures the functions of nutrition, respiration, and excretion. It is also an endocrine gland upon which the hormonal equilibrium of pregnancy depends.

Volume II deals with organogenesis, including development of the skeleton, the face, and the digestive, respiratory, urinary, genital, and circulatory systems. Development of the nervous system, the sensory organs, and endocrine glands is covered in Volume III.

We are grateful to numerous students and colleagues whose cooperation has aided preparation of this book.

THE AUTHORS

TRANSLATOR'S PREFACE

Translation of this work was undertaken in order to make available in English this excellent and unusual aid for the teaching and study of mammalian, primarily human, embryology.

This book emphasizes visual presentations. It combines the use of exceptionally clear and instructive drawings with photomicrographs and concise but complete text in an exposition of the dynamic aspects of development.

Thus, the three volumes of this book will be of help in preparation and review for students, research workers, medical practitioners such as obstetricians and pediatricians, and others who are concerned with embryology. Analysis of the precise timing of various stages of human development makes it especially useful for all who are interested in the study and prevention of congenital malformations.

The help of Kenneth Thompson in the preparation of this work is gratefully acknowledged.

LUCILLE S. HURLEY.

TABLE OF CONTENTS

PRELIMINARY DEFINITIONS

PLANES OF SECTION

Conventional planes in anatomy. — These are the classic planes of space in 3 dimensions : only the terminology is special, with the subject in an upright position facing the observer,

— *the frontal plane* (1) is parallel with the forehead;

— *the sagittal plane* (2) is perpendicular to the forehead, and parallel with the long axis of the nose;

— *the transverse plane* (3) is horizontal, parallel with an imaginary line joining the eyes.

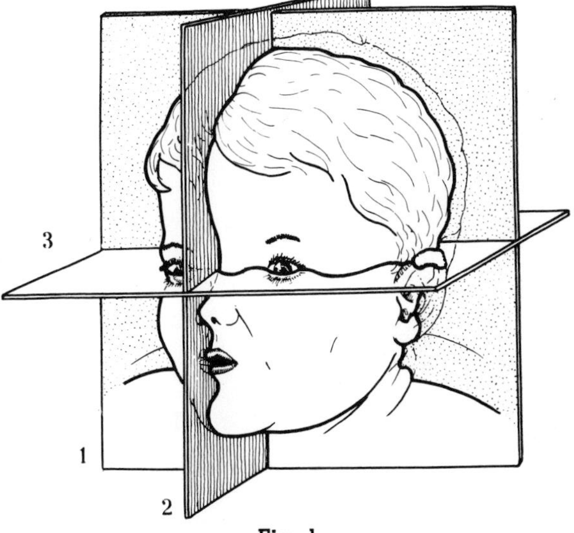

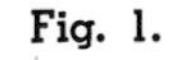
Fig. 1.

Special features in embryology. — The embryo is curved (see *Flexion,* p. 46) and a reference position has not been defined as it has in the adult. This curvature modifies the reciprocal relationships of the frontal and transverse planes : a section which is frontal at the level of the head can be transversal at the level of the trunk, as in figure 2.

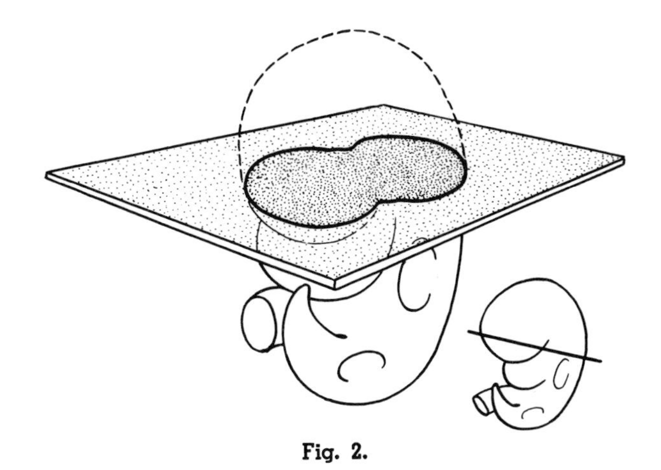

Fig. 2.

FORMATION

The gametes are formed in the gonads which also have a hormonal function.

I. — SPERMATOGENESIS

Spermatozoa are formed in the testis, from basic cells or spermatogonia. Production of spermatozoa is continuous from puberty to death.

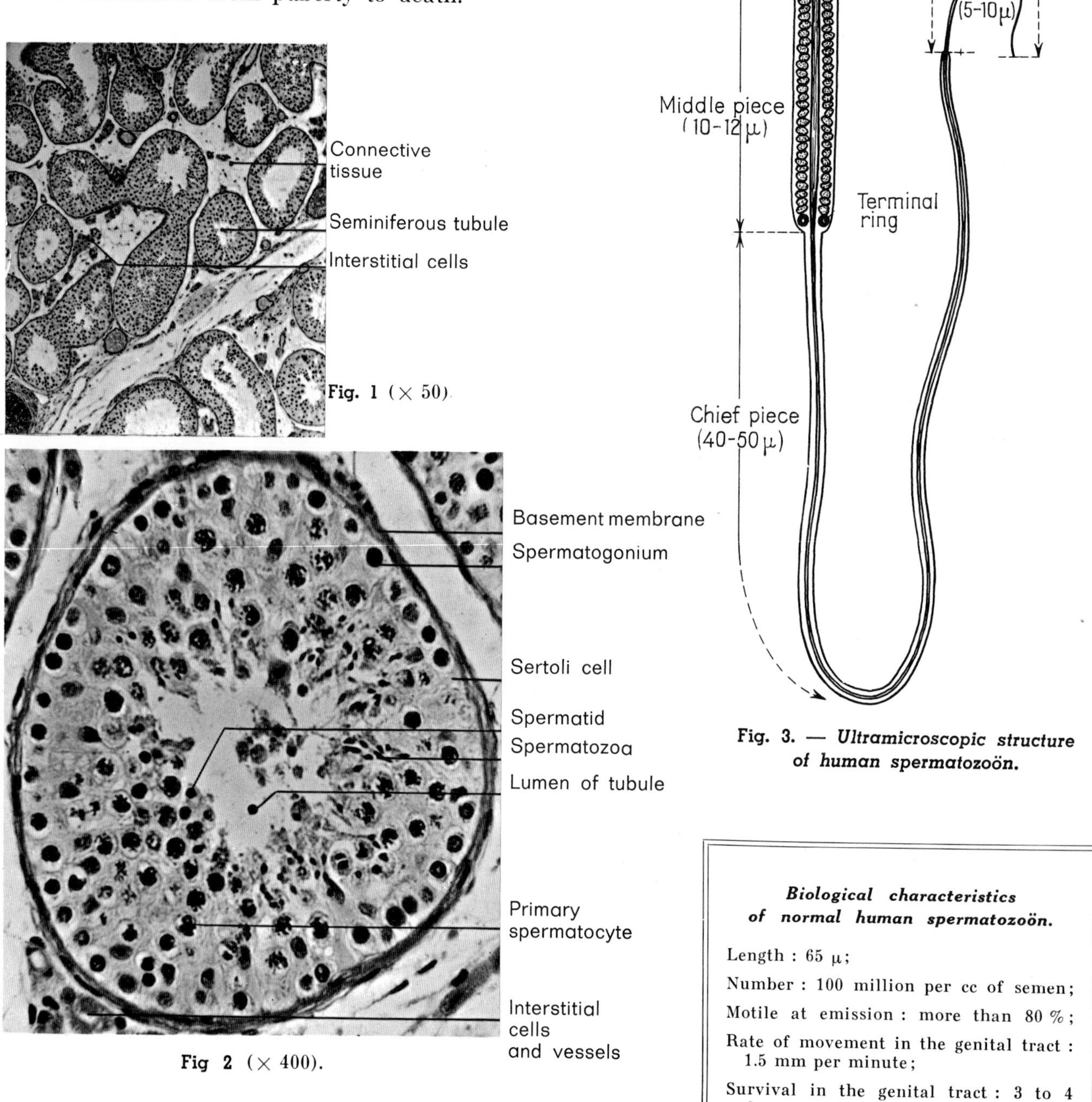

Fig. 1 (× 50)

Fig 2 (× 400).

Fig. 1 and 2. — *Formation of spermatozoa in the seminiferous tubules.* Section of human testis.

Fig. 3. — *Ultramicroscopic structure of human spermatozoön.*

Biological characteristics of normal human spermatozoön.

Length : 65 μ;

Number : 100 million per cc of semen;

Motile at emission : more than 80 %;

Rate of movement in the genital tract : 1.5 mm per minute;

Survival in the genital tract : 3 to 4 days.

OF THE GAMETES

II. — OÖGENESIS

The ova are formed in the ovary from cells called oögonia. All of the oögonia are present at birth (200 to 300,000). Of these, only 200-300 will reach maturity after puberty.

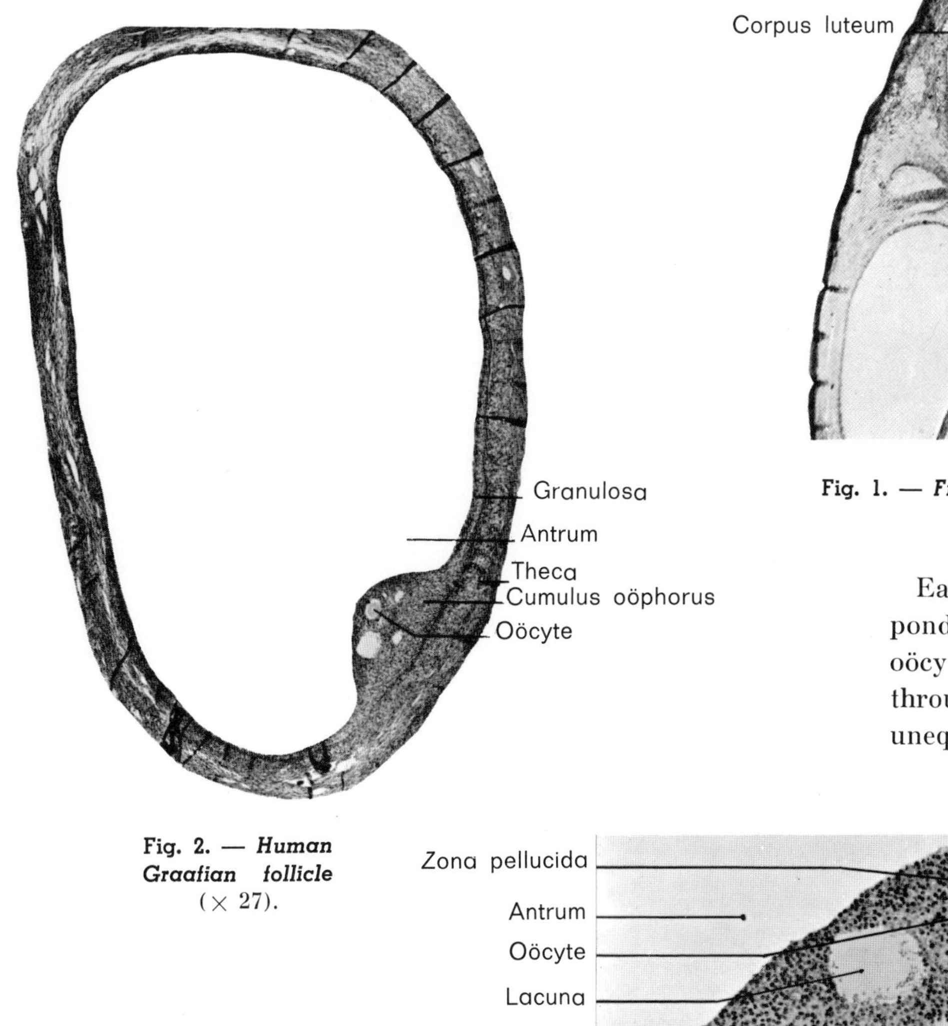

Fig. 1. — ***Fragment of human ovary*** (× 14).

Each menstrual cycle corresponds to the maturation of an oöcyte, which becomes an ovum through division, yielding cells of unequal size.

Fig. 2. — ***Human Graafian follicle*** (× 27).

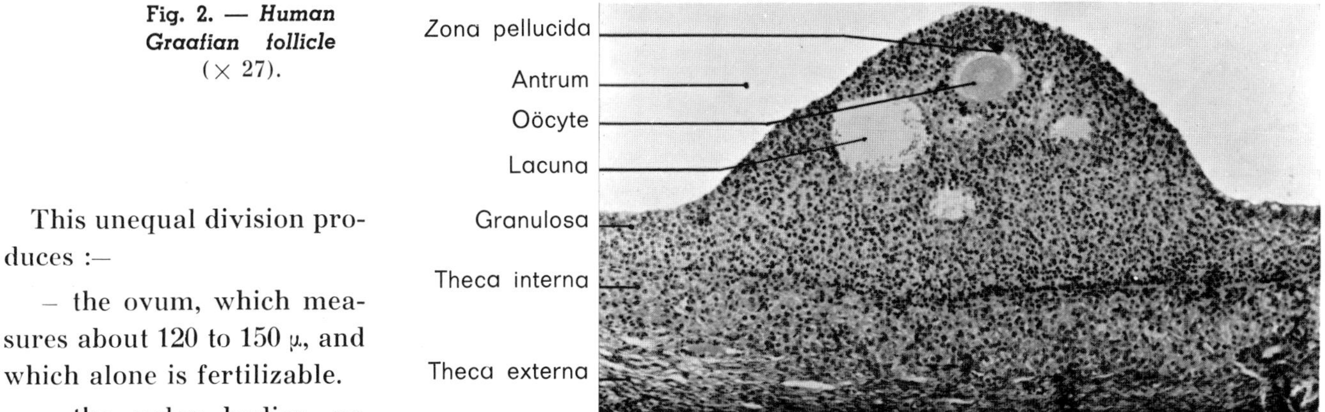

Fig. 3. — ***Cumulus oöphorus*** (× 90).

This unequal division produces :—

– the ovum, which measures about 120 to 150 μ, and which alone is fertilizable.

– the polar bodies, no larger than 10 μ.

ANOMALIES

Normal gametogenesis is a delicate process which has a double purpose.

1st. Reduction to half the number of chromosomes as well as redistribution of the hereditary material. This is accomplished by ***meiosis*** (the combination of 2 divisions involving a single synthesis of DNA and an exchange of chromosome segments).

2nd. Acquisition of special ***form*** and ***function*** by the reproductive cells which make them especially suited for fertilization.

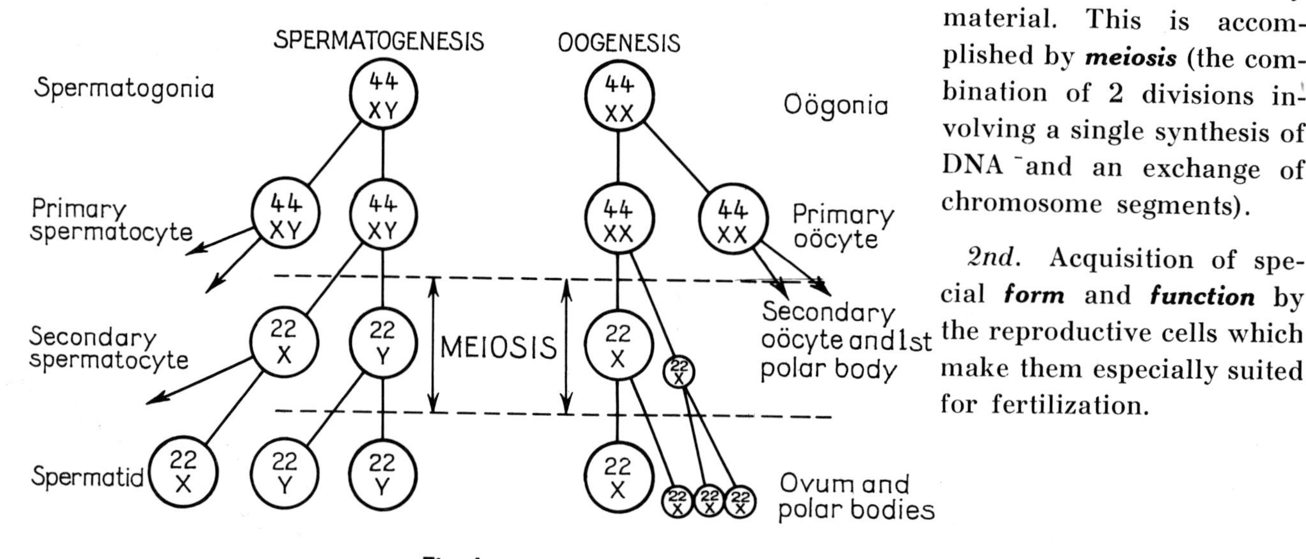

Fig. 1.

MORPHOLOGICAL ANOMALIES

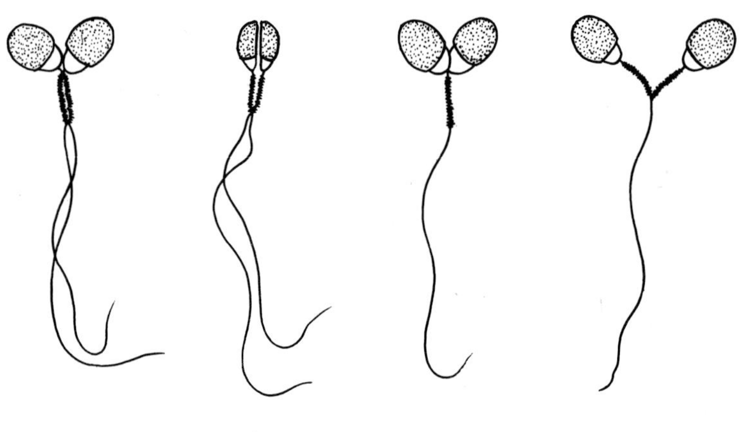

Fig. 2. — *Examples of abnormal spermatozoa.*

These types are seen even in normal semen, but usually do not exceed 20 %.

The double sperm forms may be due to failure of disjunction during spermatogenesis.

Follicular cells

Ovarian stroma

a *b*

Fig. 3. — *Unusual cell types in a fetal ovary.*

a) Oöcyte with 2 nuclei.

b) Two oöcytes in the same follicle.

OF GAMETOGENESIS

CHROMOSOMAL ANOMALIES

During meiosis abnormalities can occur in distribution of the chromosomal material between the gametes.

1. ***Anomalies involving the autosomes*** (somatic chromosomes).

Meiotic division includes a stage of chromosome pairing. This phenomenon provides the possibility of non-disjunction of one pair resulting in the formation of 2 abnormal gametes : one abnormal gamete has both chromosomes of this pair, the other has this pair of chromosomes completely missing.

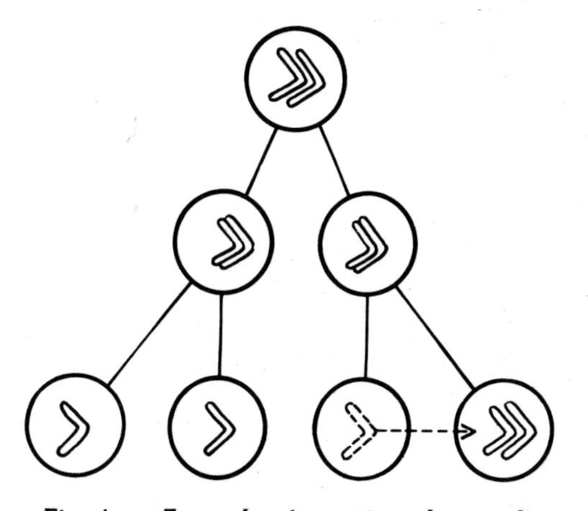

Fig. 4. — ***Example of partition abnormality in an autosomal pair :*** only the chromosomal pair concerned is shown. Of the 4 cells arising from the primordial germ cell, 2 are normal, 2 are pathological.

2. ***Abnormalities involving the sex chromosomes.***

The same type of abnormalities as described under 1. can be seen here : certain cells have no sex chromosome, others possess two (or sometimes even more) (fig. 5).

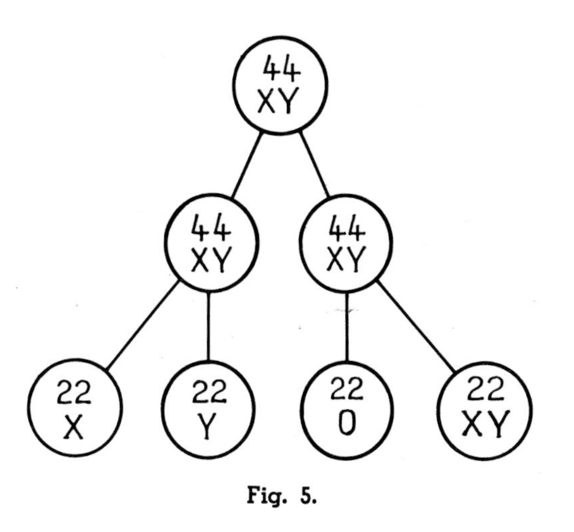

Fig. 5.

3. Chromosome abnormalities particularly affect female gametes : the greater vulnerability of female as compared with male gametes is due to sex differences in the chronology of maturation, despite identical fundamental mechanisms (see p. 6 and 7).

CHRONOLOGY

IN THE MALE

The fundamental difference apparent from these diagrams is the unequal duration of meiosis in the two sexes.

In the male, meiosis comes about within several days.

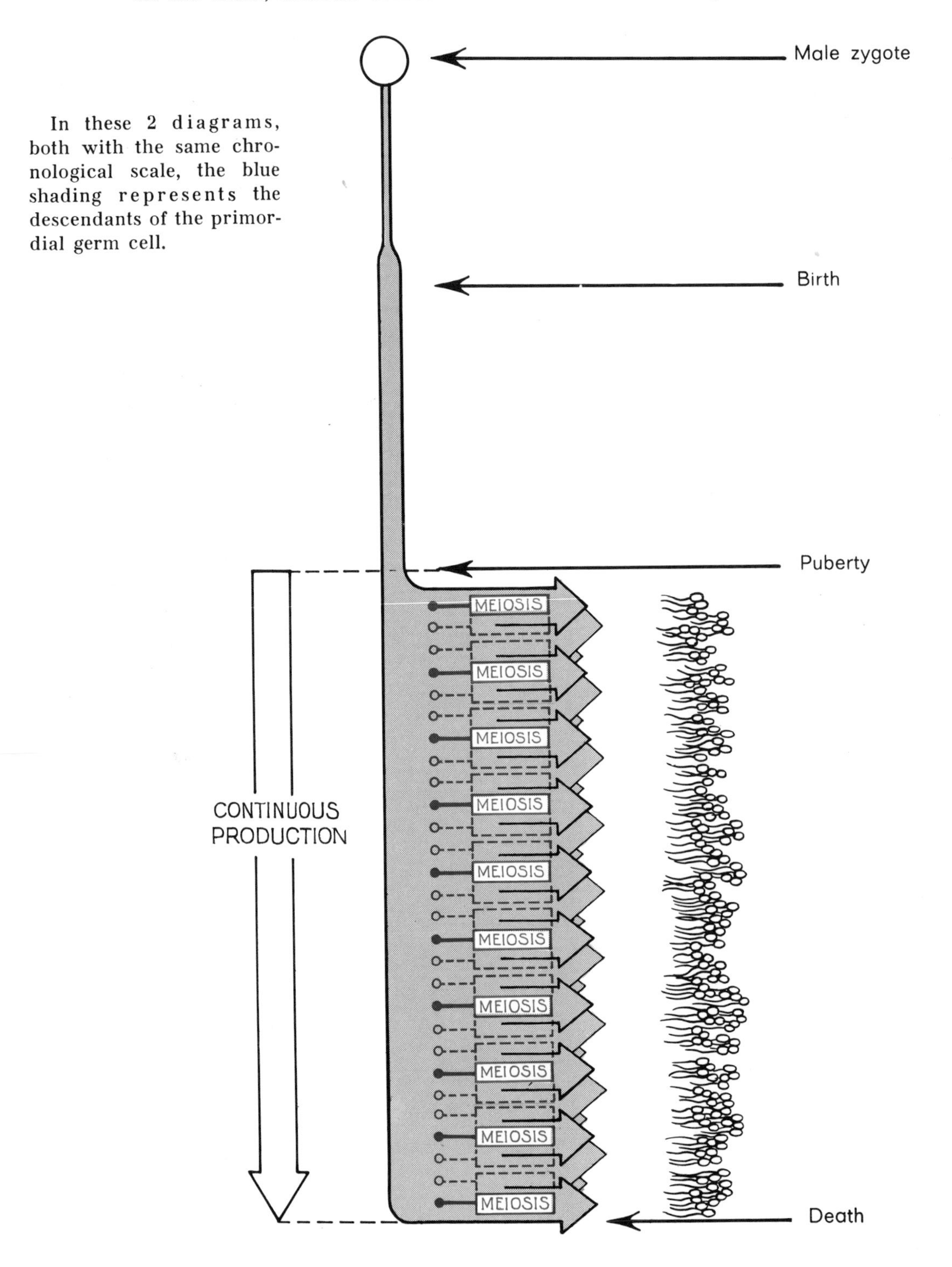

In these 2 diagrams, both with the same chronological scale, the blue shading represents the descendants of the primordial germ cell.

OF GAMETOGENESIS

IN THE FEMALE

In the female, the process begun during fetal life is suspended for a considerable time, indeed for about a dozen years.

Eventually, this delay may be a cause of chromosomal abnormalities.

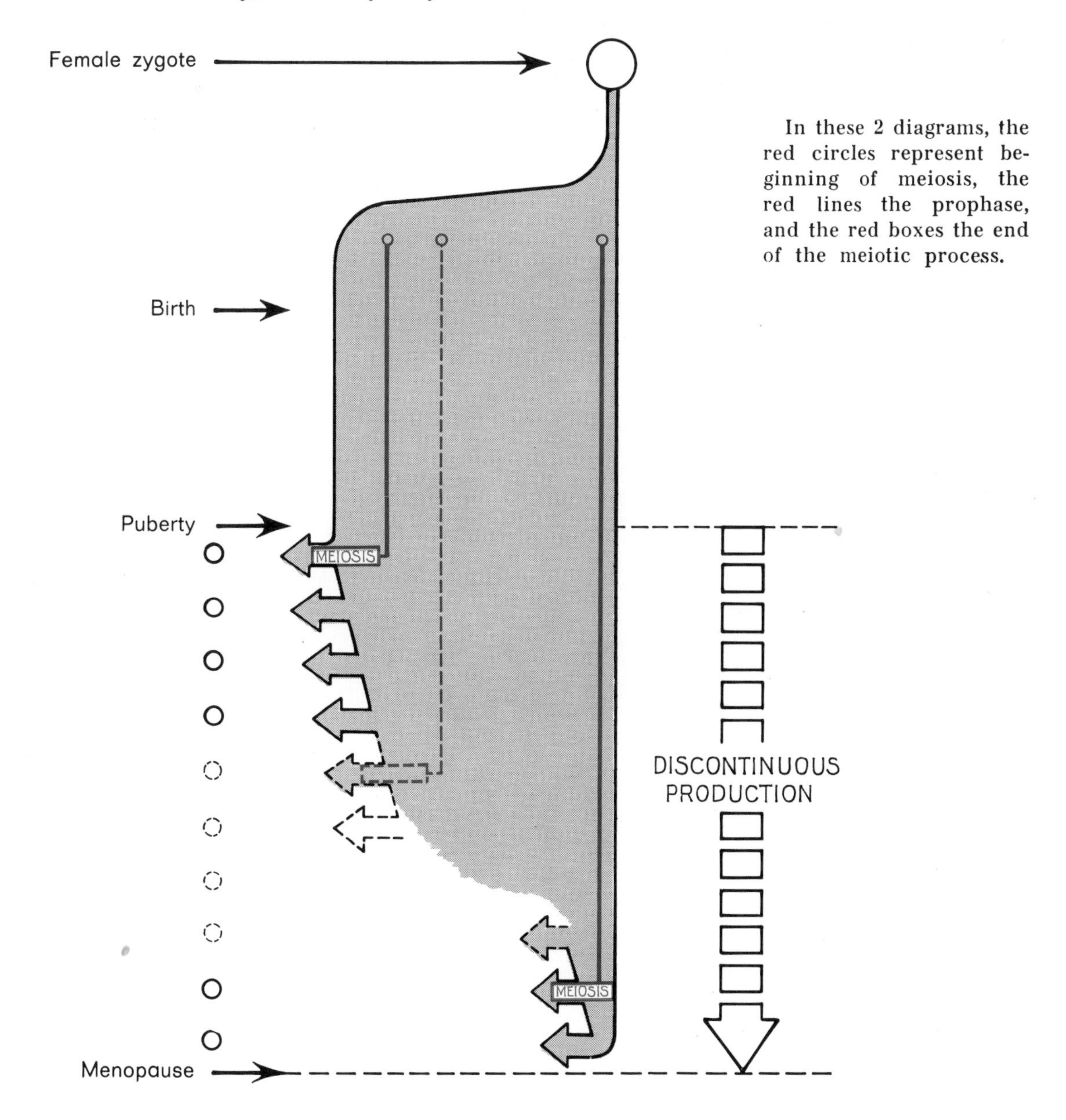

In these 2 diagrams, the red circles represent beginning of meiosis, the red lines the prophase, and the red boxes the end of the meiotic process.

PHENOMENA

Fertilization is the union of male and female gametes. It marks the beginning of pregnancy.

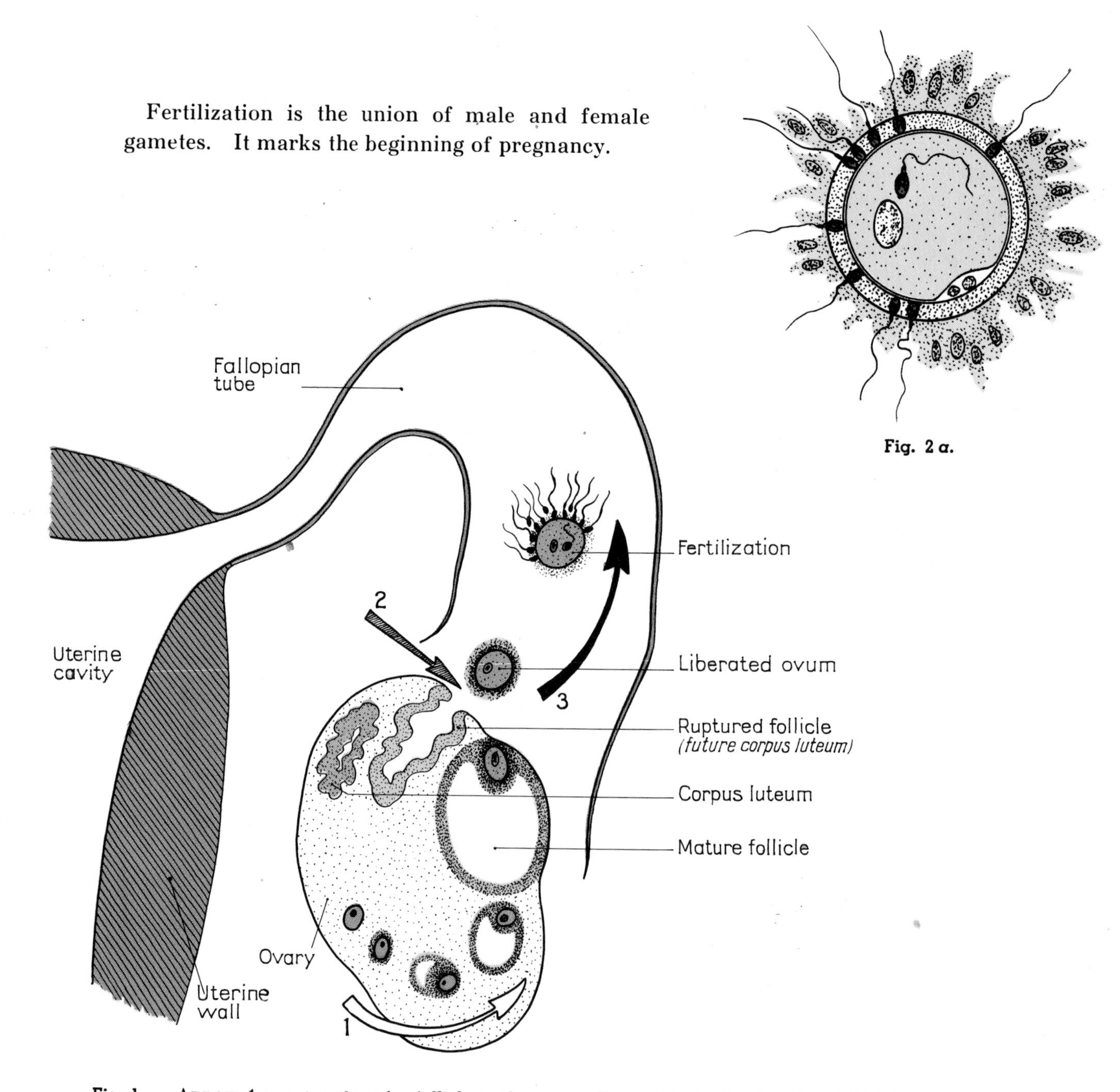

Fig. 2 *a*.

Fig. 1. — Arrow 1 : ***maturation of a follicle in the ovary*** (from oöcyte to Graafian follicle).

Arrow 2 : **ovulation :** coincides with the first maturation division and with elimination of the first polar body.

The ovum is captured by the ampulla of the Fallopian tube whose fimbriae sweep over the ovary.

Arrow 3 : ***fertilization :*** takes place in the distal third of the Fallopian tube. Spermatozoa arrive there about 10 hours after coitus. The ovum must be fertilized within 24 hours after ovulation.

OF FERTILIZATION

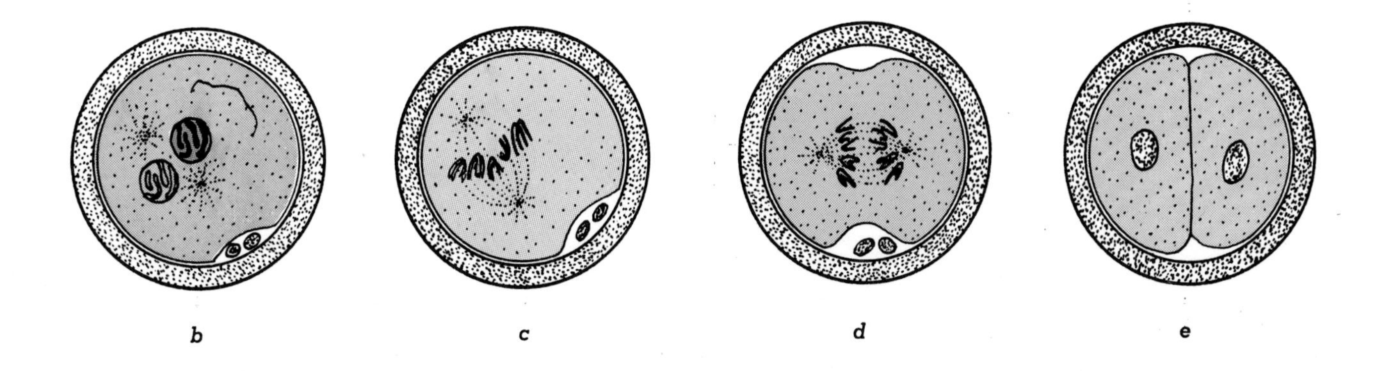

Fig. 2. — *Morphological changes in fertilization.*

a) The sperm has just penetrated the ovum; the second polar body is extruded (the division of the first polar body is not shown here).

b) Formation of the two pronuclei.

c) Metaphase of the first cleavage mitosis : the normal chromosome stock is reconstituted.

d) Anaphase of the first cleavage mitosis.

e) The first two blastomeres, still surrounded by the zona pellucida.

Consequences of fertilization.

— *Activation of the ovum.*

— *Modification of the cytoplasm and of the membrane.*

— *Modification of the nucleus* (fig. 3) :—

— reconstitution of the diploid number of chromosomes (see Karyotype, p. 98).

— determination of sex by the X or Y chromosome of the sperm.

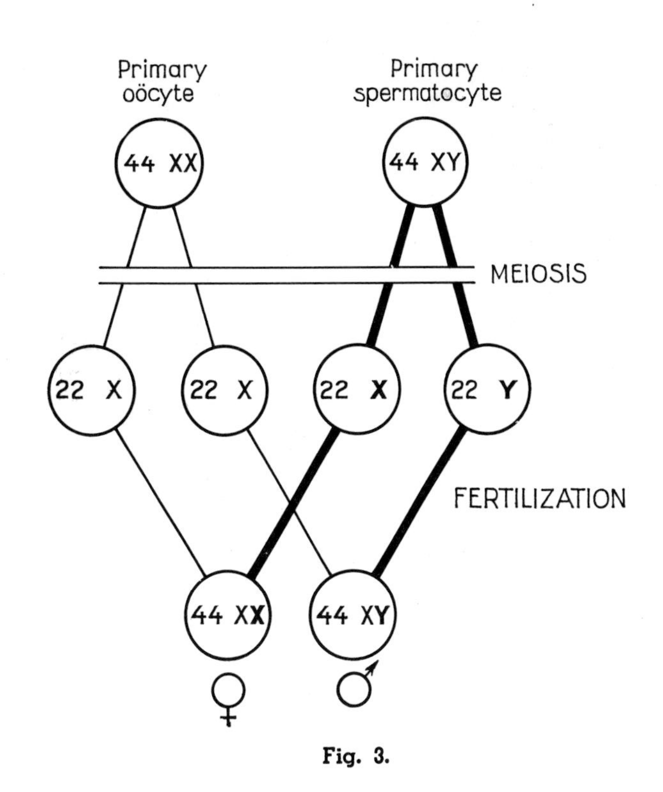

Fig. 3.

PREPARATION

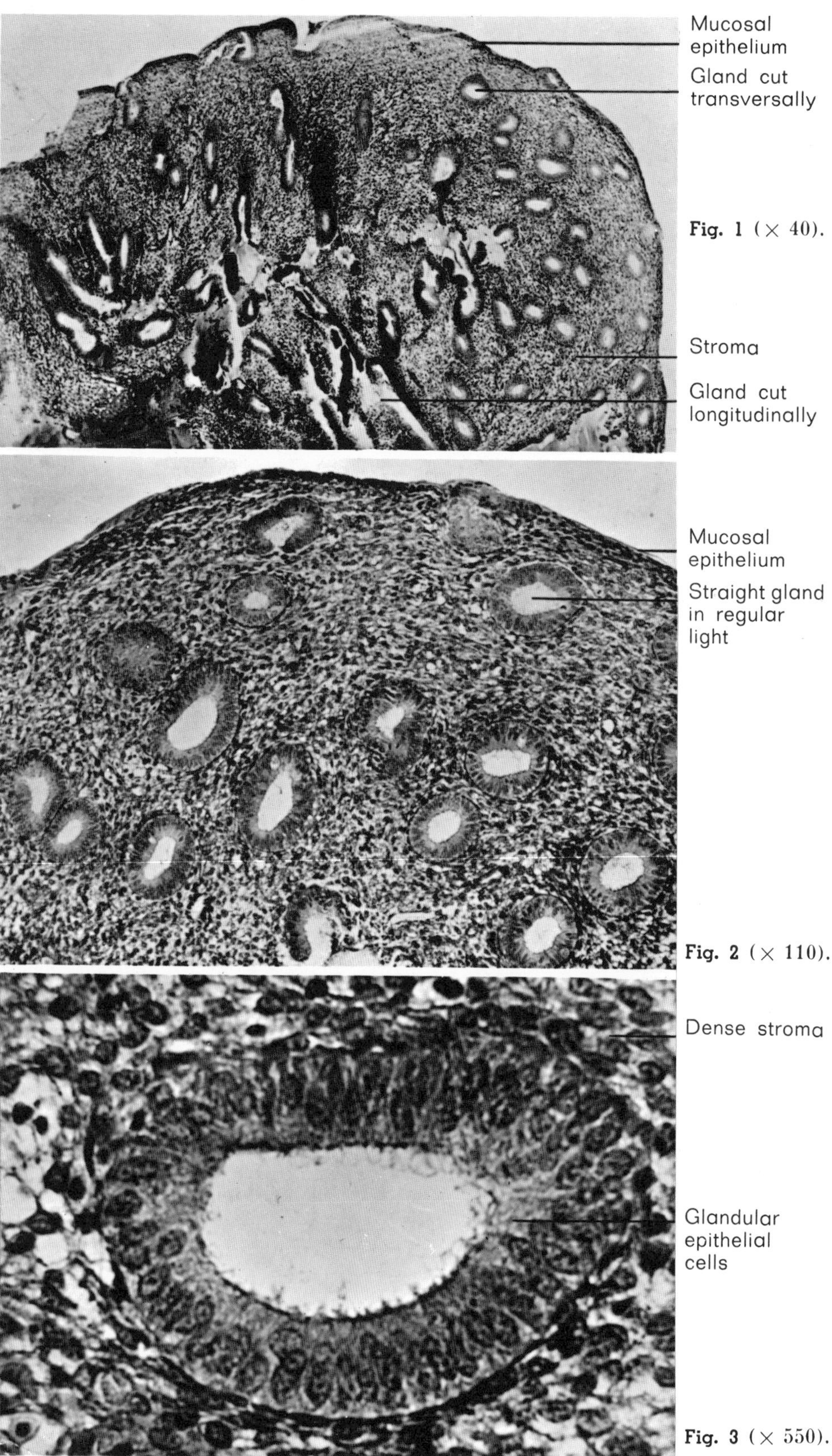

Fig. 1 (× 40).

Fig. 2 (× 110).

Fig. 3 (× 550).

Follicular phase.

I. — CHANGES IN

Implantation generally occurs on the twenty-first day of the menstrual cycle, that is, during the progestational phase. At this time the mucosa is thick, richly vascular-

Fig. 1, 2 and 3. — *Biopsy sections of human endometrium :* uterine mucosa in ***follicular or estrogenic phase.***

Predominance of proliferative changes.

FOR IMPLANTATION

THE UTERINE MUCOSA

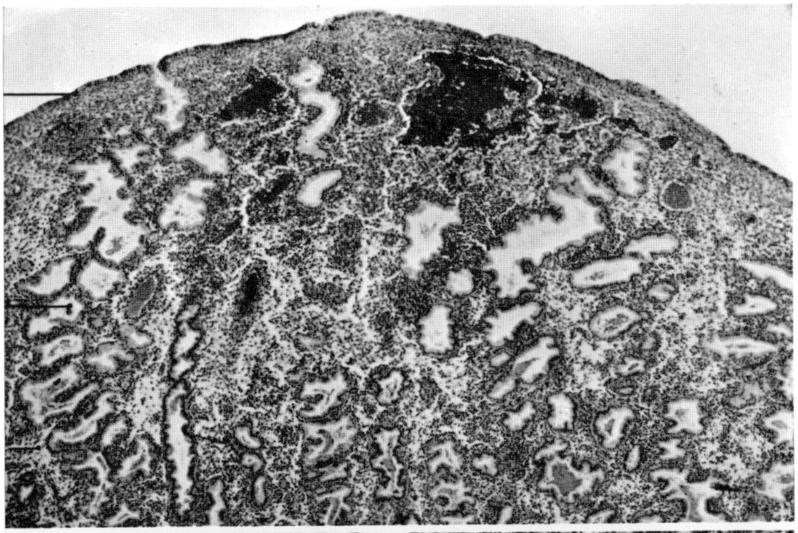
Mucosal epithelium
Convoluted gland
Stroma

Fig. 4 (× 40).

ized, and provided with large amounts of glycogen. The blastocyst thus finds conditions particularly favorable for its implantation, especially for its nutrition.

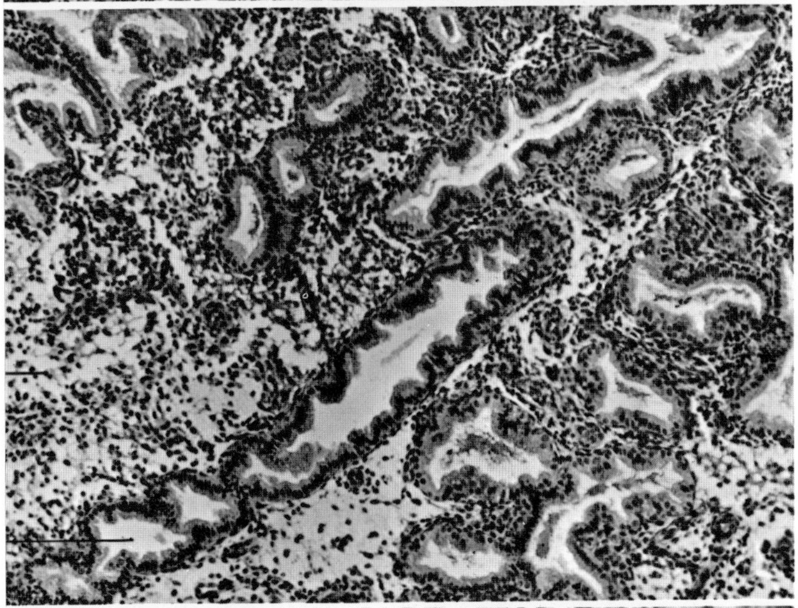
Edematous stroma
Convoluted gland

Fig. 5 (× 110).

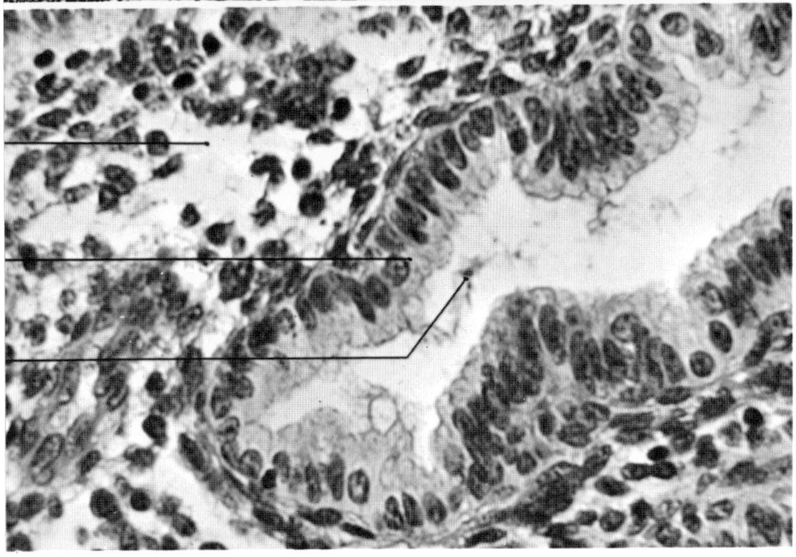
Edematous stroma
Glandular epithelial cells
Secreted material, rich in glycogen

Fig. 6 (× 550).

Fig. 4, 5 and 6. — ***Biopsy sections of human endometrium :*** uterine mucosa in ***progestational phase.***

Proliferation and predominance of secretion, congestion, and edema.

Progestational phase.

II. — HORMONAL

ACTION OF OVARIAN HORMONES ON THE ENDOMETRIUM

During the course of each menstrual cycle, the uterine mucosa undergoes preparation for implantation which is directly conditioned by the ovarian hormones estrogen (fig. 1) and progesterone (fig. 2).

The endometrium is shed during menstrual bleeding, but immediately begins regenerating. From the 4th day on, regeneration is accelerated. Proliferative processes are predominant during the follicular phase, and also continue during the progestational phase when secretory phenomena prevail.

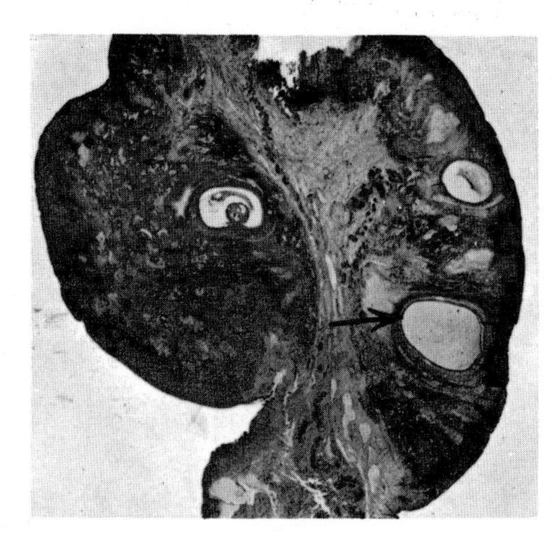

Fig. 1.

Fig. 1. — ***Graafian follicle,*** principal source of ***estrogen*** (from the theca interna).

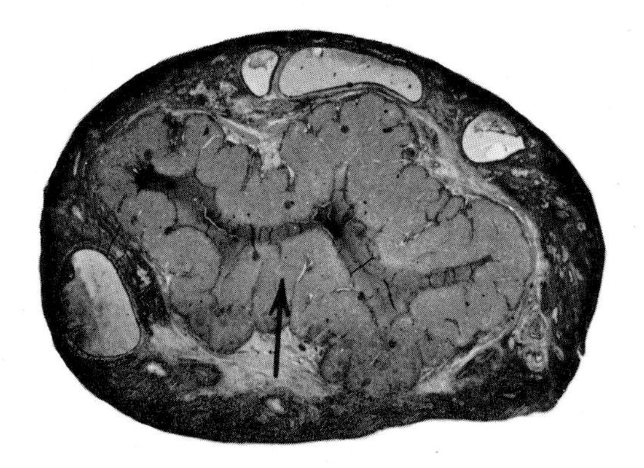

Fig. 2.

Fig. 2. — ***Corpus luteum of pregnancy,*** principal source of **progesterone.**

Follicular phase.

Progestational phase.

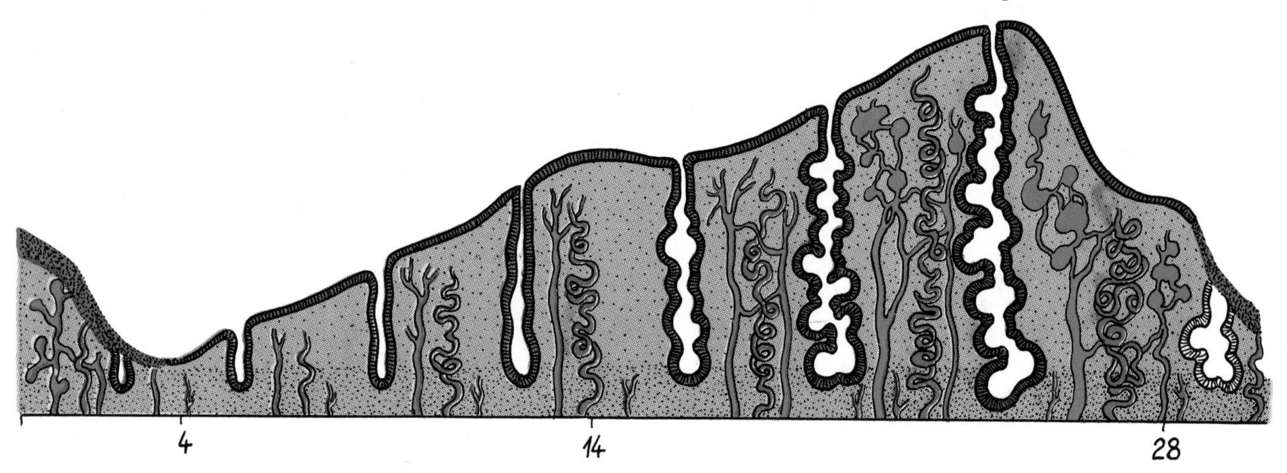

Fig. 3. — ***Morphological changes in the uterine mucosa during the menstrual cycle :*** proliferation of the endometrium involves not only the epithelium, the glands, and the stroma, but also and in a very important way, the blood vessels.

ASPECTS

HYPOPHYSEAL-OVARIAN CORRELATIONS

Hypophysis. — Endocrine activity of the ovary is under the control of the anterior lobe of the pituitary which, in human beings, secretes two gonad-stimulating hormones (gonadotropes or gonadotrophins) (fig. 4).

Follicle-stimulating hormone (FSH) is elaborated from the beginning of the cycle. It determines growth of the follicle.

Luteinizing hormone (LH) is secreted in the middle of the cycle. It acts synergistically with FSH to provoke ovulation. It stimulates the development of the corpus luteum.

Ovary. — Under the influence of the hypophyseal gonadotrophins, endocrine activity of the ovary is diphasic : secretion of *estrogen* during the first phase, then of *estrogen and progesterone* during the second phase (fig. 4). Secretion of progesterone is detectable even before formation of the corpus luteum.

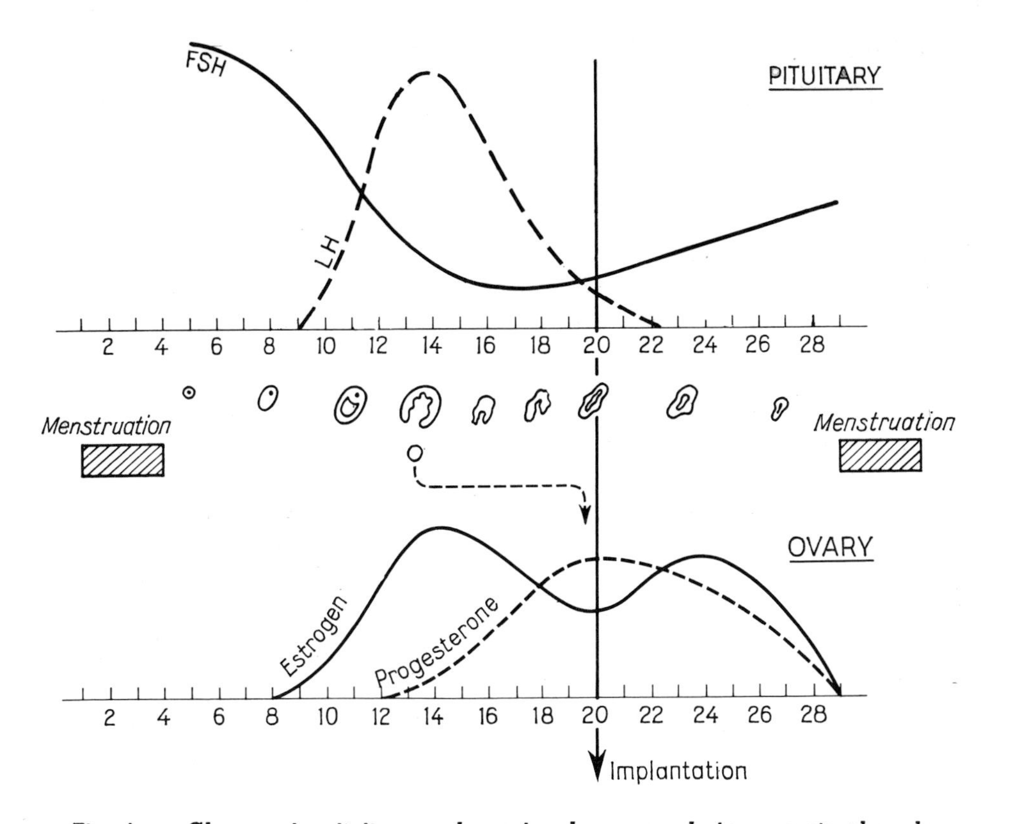

Fig. 4. — *Changes in pituitary and ovarian hormones during menstrual cycle.*

The vertical arrow indicates the time when implantation would take place if fertilization had occurred.

n

FIRST WEEK

I. — CLEAVAGE

The fertilized ovum, through cleavage, reaches the *morula* stage. Then, forming a central cavity, it becomes a *blastocyst,* which implants itself in the uterine mucosa on the 6th day.

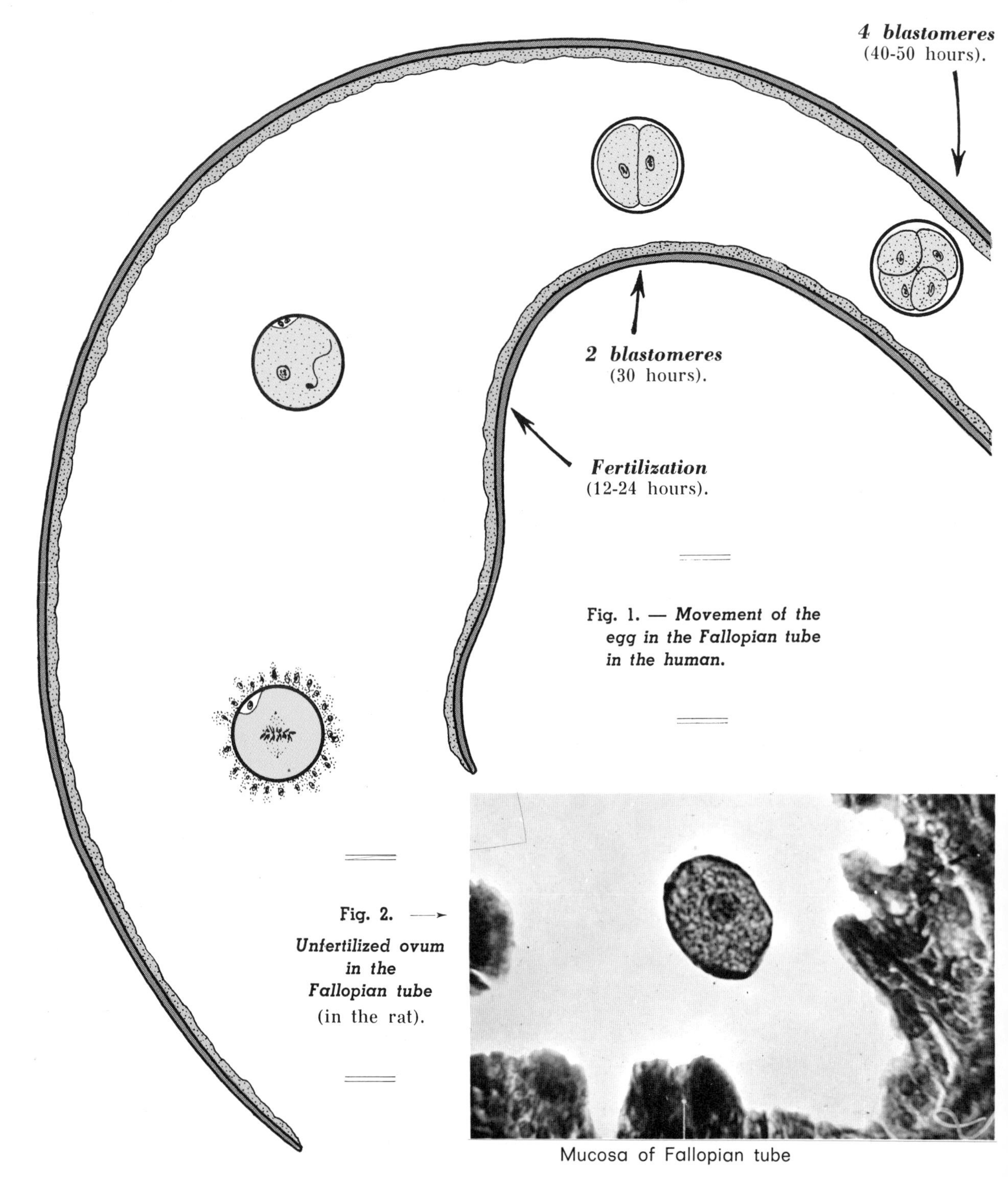

Fig. 1. — *Movement of the egg in the Fallopian tube in the human.*

Fig. 2. → ***Unfertilized ovum in the Fallopian tube*** (in the rat).

OF DEVELOPMENT

During passage through the tube, until the end of the morula stage, the egg undergoes practically no change in volume (150 μ). It remains surrounded by the zona pellucida which it loses upon entering the uterus. It progresses under the influence of peristaltic movements of the Fallopian tube and of ciliary movements of the tubular epithelium.

During this phase, the egg lives on its reserves (but they are reduced, it is an *alecithinic* egg) and on tubular secretions.

Survival of the egg and its transport in the genital tract, as well as implantation of the blastocyst, depend on the hormonal secretions of the ovary and the anterior pituitary.

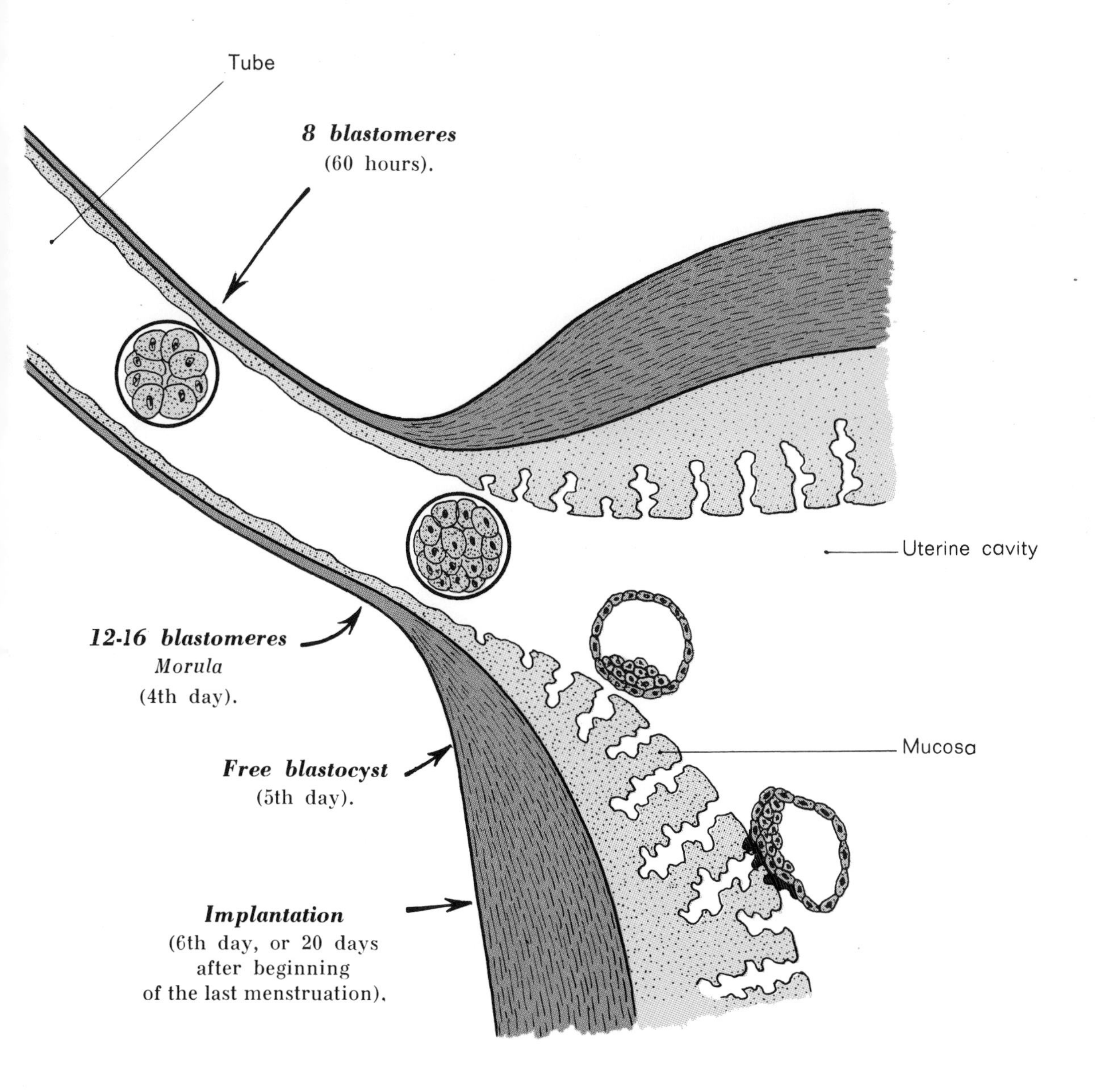

II. — IMPLANTATION

The blastocyst begins to implant about the 6th or 7th day.

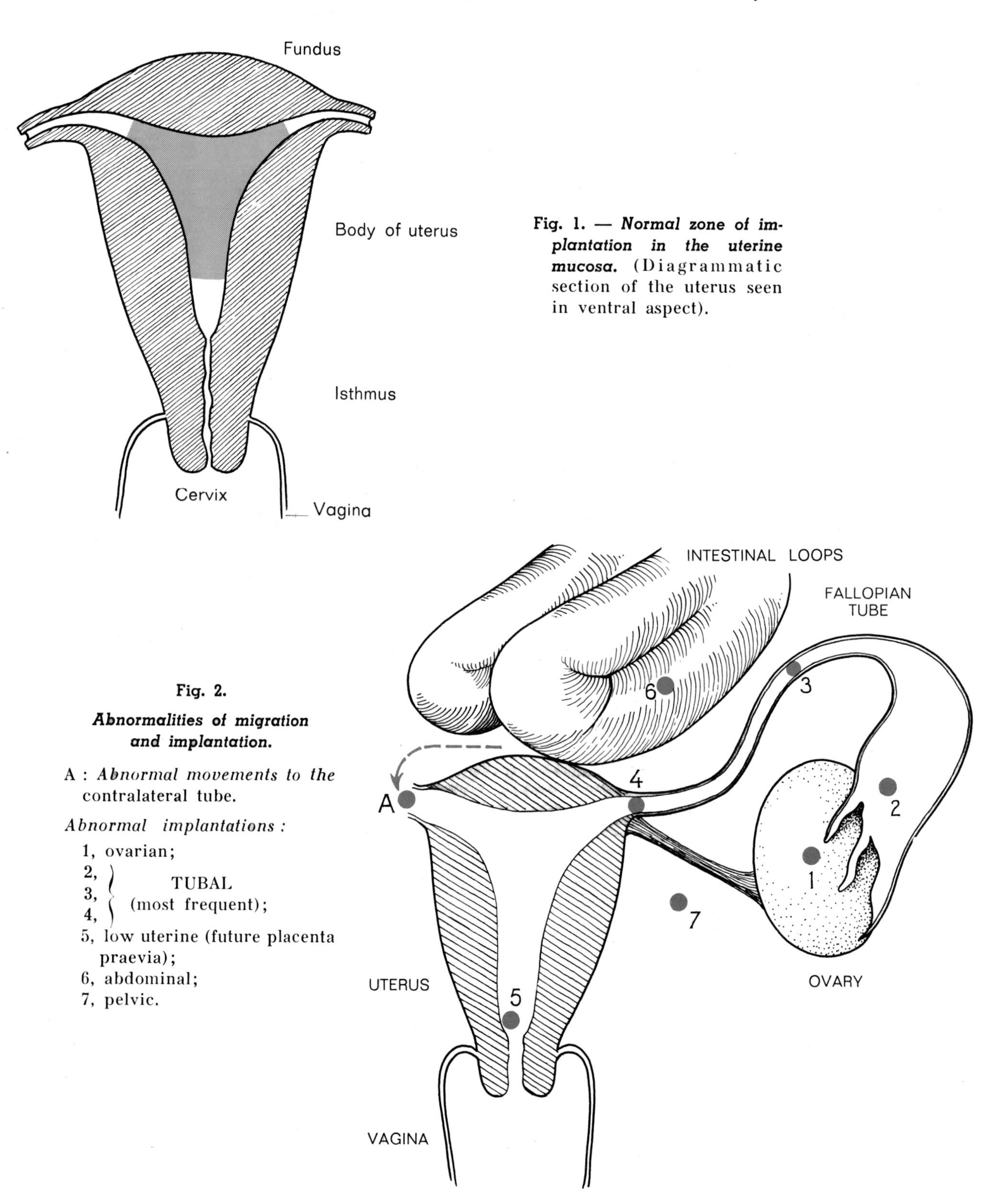

Fig. 1. — *Normal zone of implantation in the uterine mucosa.* (Diagrammatic section of the uterus seen in ventral aspect).

Fig. 2.

Abnormalities of migration and implantation.

A : *Abnormal movements to the contralateral tube.*

Abnormal implantations :

1, ovarian;
2, 3, 4, TUBAL (most frequent);
5, low uterine (future placenta praevia);
6, abdominal;
7, pelvic.

The fertilized ovum is implanted at the embryonic pole due to the lytic activity of the syncytiotrophoblast. At this stage, the embryoblast begins to differentiate clearly from the trophoblast.

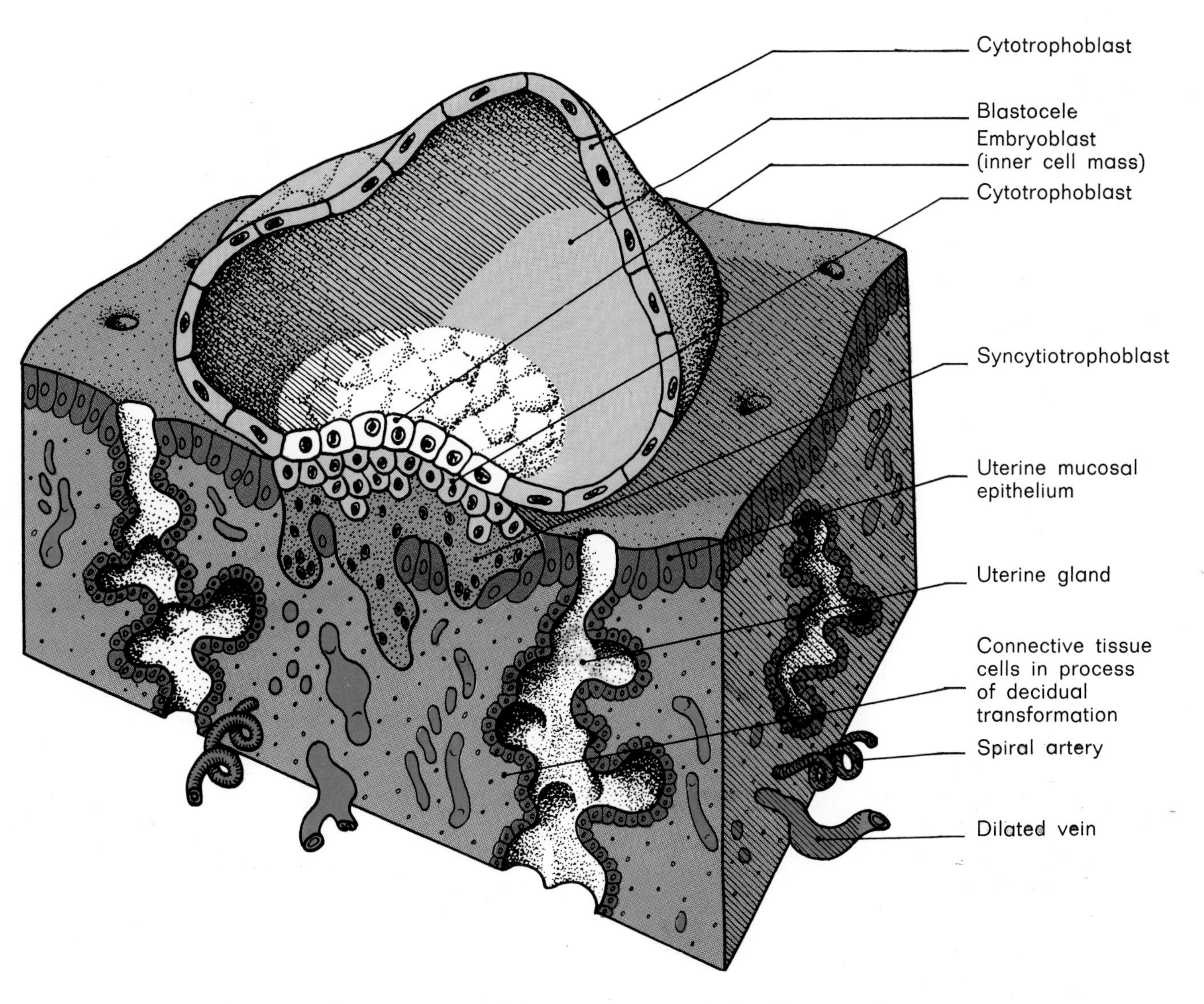

Fig. 3. — ***Diagrammatic relief representation of the blastocyst in process of implantation in the uterine mucosa.***

At the time of implantation, the uterine mucosa is at the 21st day of the cycle. The rich vascularization, edema, and secretion of mucus and glycogen favor implantation of the blastocyst.

SECOND WEEK : FORMATION

At the end of the first week, the blastocyst is in the embryoblast stage (fig. 1). During the second week, the ***entoderm*** (fig. 1), then the

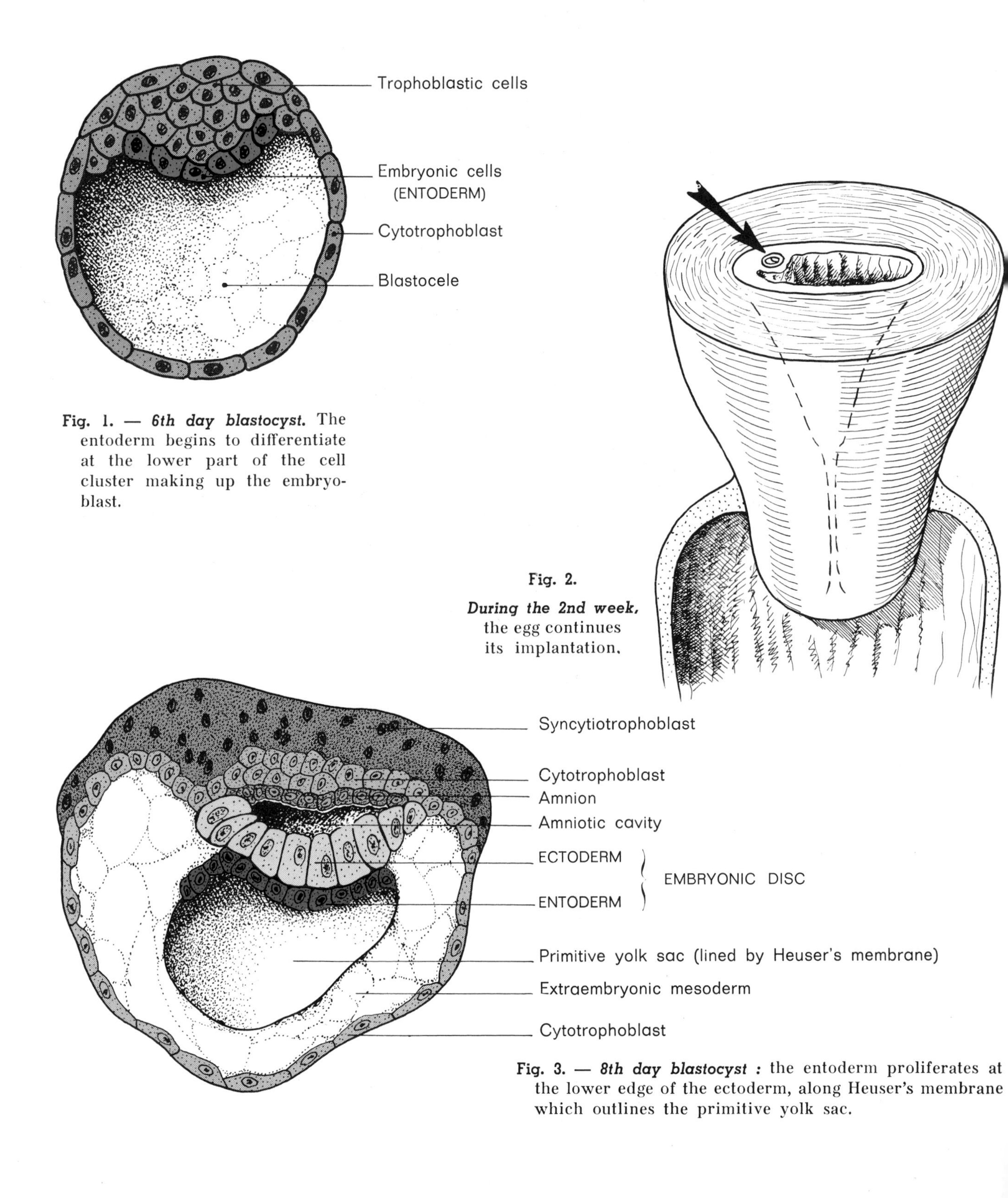

Fig. 1. — ***6th day blastocyst.*** The entoderm begins to differentiate at the lower part of the cell cluster making up the embryoblast.

Fig. 2.
During the 2nd week, the egg continues its implantation.

Fig. 3. — ***8th day blastocyst :*** the entoderm proliferates at the lower edge of the ectoderm, along Heuser's membrane which outlines the primitive yolk sac.

ectoderm (fig. 3) are differentiated successively from the lower part of this cellular mass, while the egg buries itself in the uterine mucosa (fig. 2).

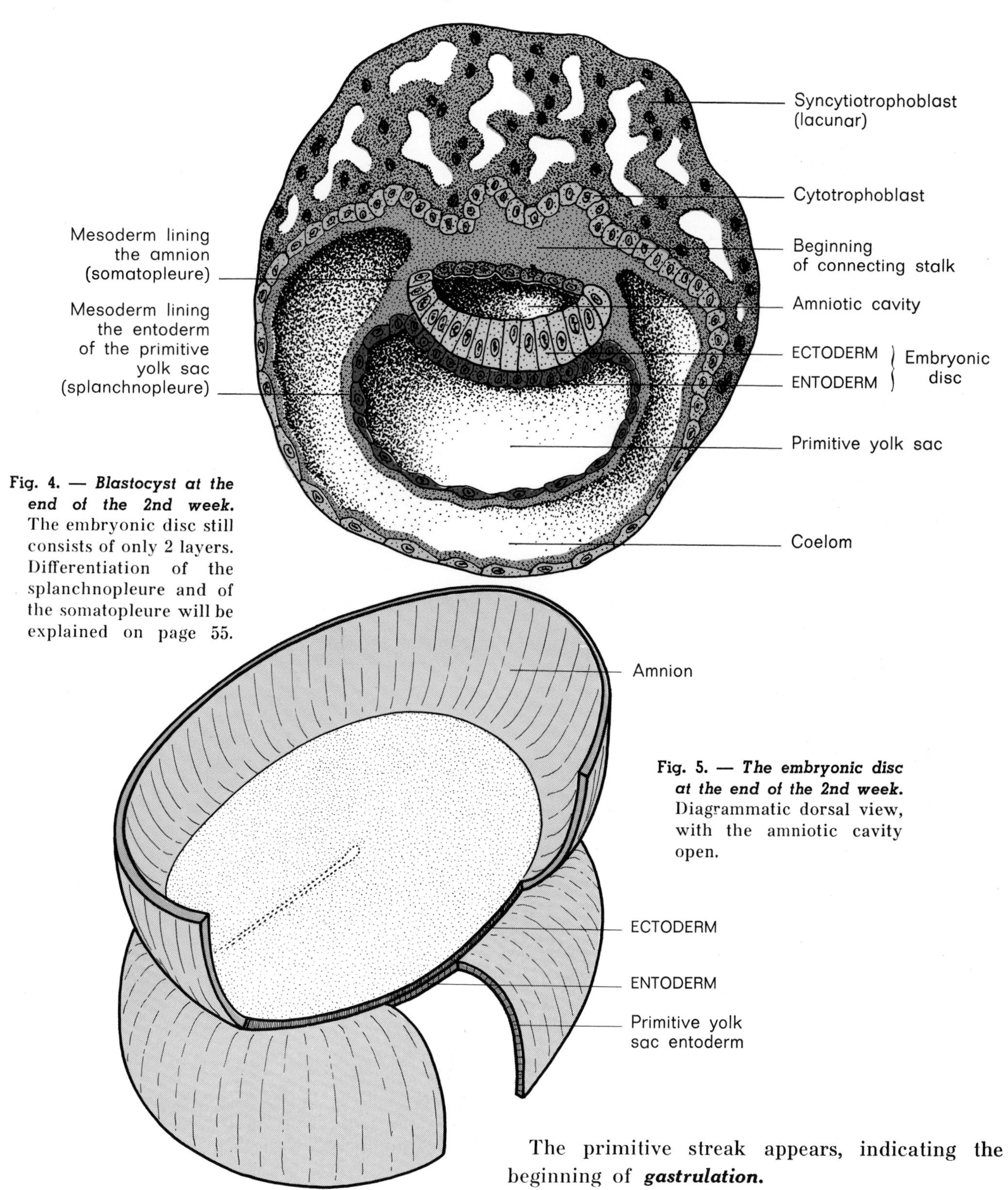

Fig. 4. — *Blastocyst at the end of the 2nd week.* The embryonic disc still consists of only 2 layers. Differentiation of the splanchnopleure and of the somatopleure will be explained on page 55.

Fig. 5. — *The embryonic disc at the end of the 2nd week.* Diagrammatic dorsal view, with the amniotic cavity open.

The primitive streak appears, indicating the beginning of ***gastrulation.***

THIRD WEEK :

At the beginning of the 3rd week, an important process occurs : ***gastrulation,*** the formation of the third layer of the embryo, the *mesoderm.*

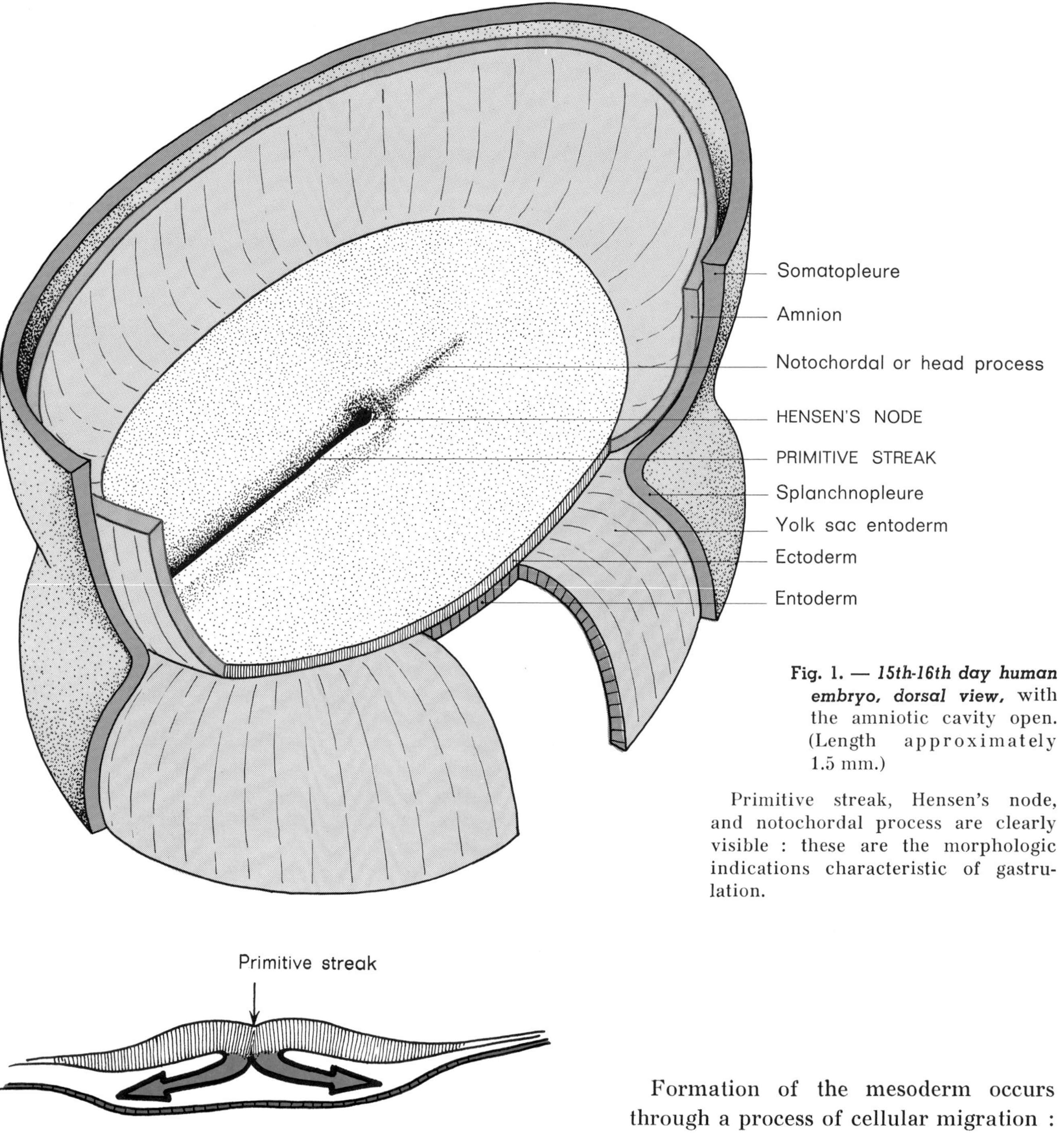

Fig. 1. — *15th-16th day human embryo, dorsal view,* with the amniotic cavity open. (Length approximately 1.5 mm.)

Primitive streak, Hensen's node, and notochordal process are clearly visible : these are the morphologic indications characteristic of gastrulation.

Fig. 2. — Cellular movements at the level of the primitive streak. *(Cross section.)*

Formation of the mesoderm occurs through a process of cellular migration : *ectodermal cells glide downward at the level of the primitive streak* (fig. 2).

GASTRULATION

I. — MOVEMENTS IN GASTRULATION

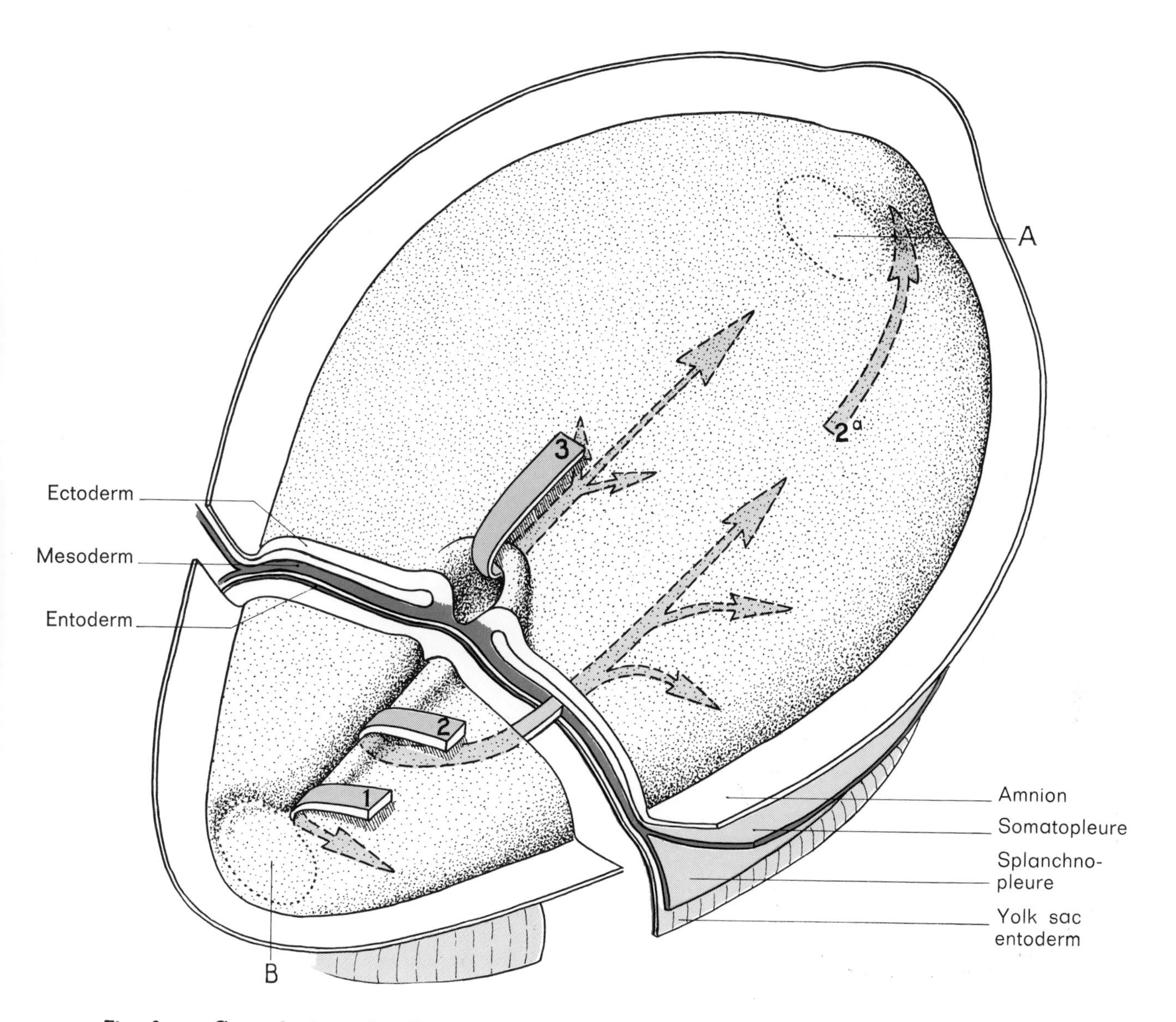

Fig. 3. — *General view of cell migration at the time of gastrulation.* The arrows show the direction of ectodermal cell movements :

— *arrow 1* : origin of mesoderm of caudal end;

— *arrow 2* : origin of lateral mesoderm;

— *arrow 2* a : part of the lateral mesoderm reaches the cephalic end;

— *arrow 3* : origin of notochordal substance (invaginated at level of Hensen's node).

Letters A and B indicate two regions where mesoderm is not interposed between ectoderm and entoderm : these are the future pharyngeal (A) and cloacal (B) membranes.

The posterior third of the embryo was cross sectioned to show the relationship of the three layers.

II. — FORMATION

The relatively simple formation of the lateral mesoderm (by invagination at the primitive streak level), contrasts with the complexity of the processes leading to the development of the notochord.

Three stages can be distinguished : notochordal process, prochordal plate, and notochord.

During this development, the region of the notochordal process, that is, the entire area cephalic to the primitive streak, undergoes preferential growth, and Hensen's node seems to recede toward the caudal end.

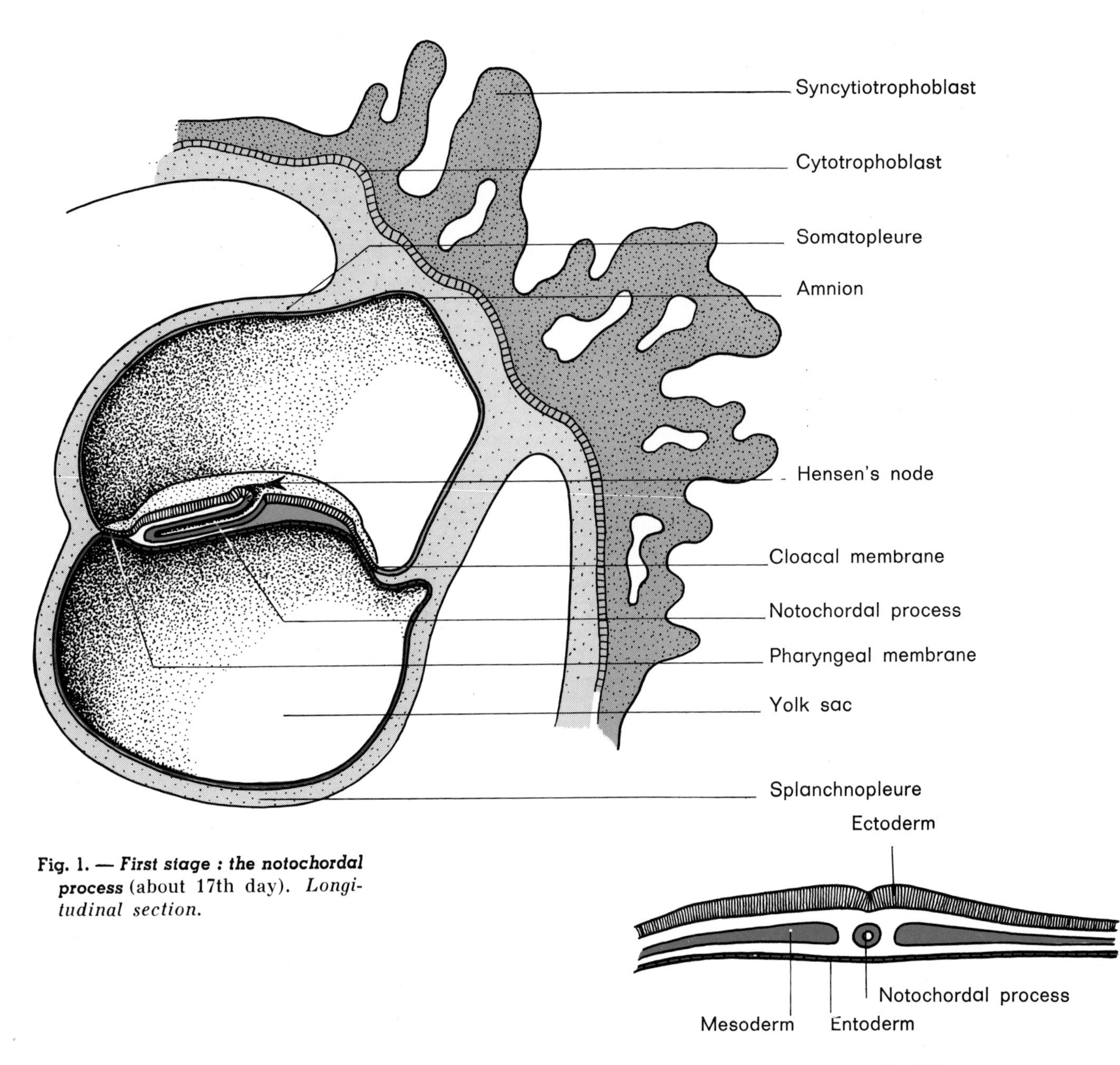

Fig. 1. — *First stage : the notochordal process* (about 17th day). *Longitudinal section.*

Fig. 1 *a*. — *Notochordal process.*
Cross section.

OF THE NOTOCHORD

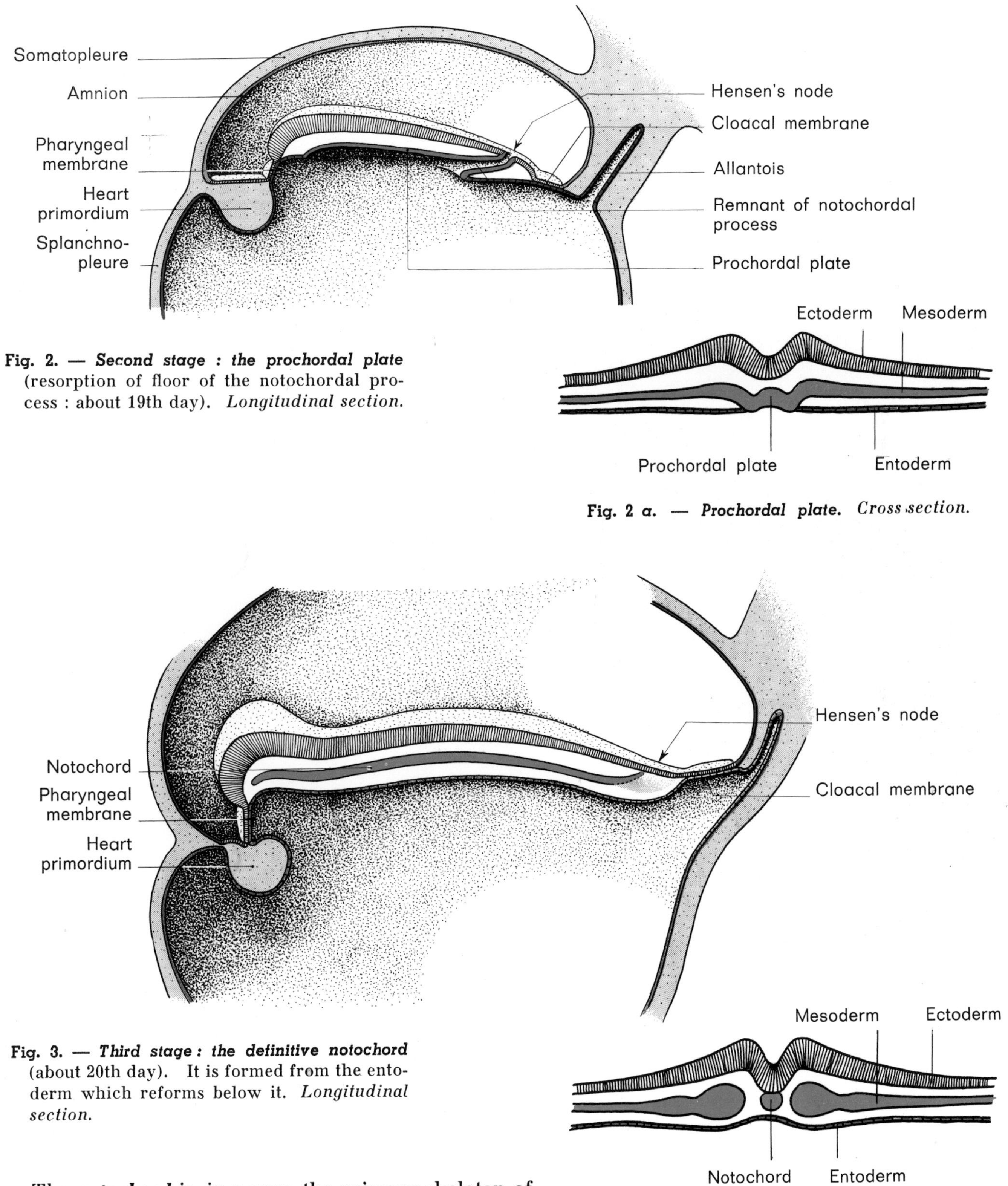

Fig. 2. — *Second stage : the prochordal plate* (resorption of floor of the notochordal process : about 19th day). *Longitudinal section.*

Fig. 2 a. — *Prochordal plate.* *Cross section.*

Fig. 3. — *Third stage : the definitive notochord* (about 20th day). It is formed from the entoderm which reforms below it. *Longitudinal section.*

Fig. 3 a. — *Definitive notochord.* *Cross section.*

The ***notochord*** is, in a way, the primary skeleton of the 3-layer embryo.

III. — MORPHOLOGIC ASPECT

For laboratory study of gastrulation, it is convenient to use the ***rabbit,*** in which formation of the mesodermal layer is very similar to that in humans.

Ovulation in the rabbit is provoked by coitus : release of the ovum occurs 6 hours later.

Gestation lasts 30 days. Gastrulation takes place 8 days after fertilization.

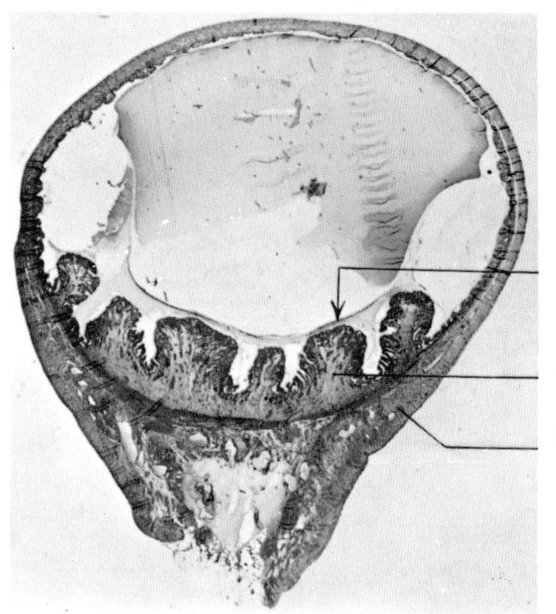

Fig. 1. — *Uterine horn of rabbit at 8th day of gestation.* *Cross section* of whole circumference (× 8).

The blastocyst is fixed to the surface of the mucosa. The cells surrounding the blastocyst appear as a dark continuous line at this magnification.

The embryo can be found on this line opposite the thickening of the maternal mucosa (see also fig. 2).

This type of implantation, in which the blastocyst remains at the surface of the maternal mucosa, is called *central.*

In humans, implantation is *interstitial,* for the blastocyst burrows into the uterine mucosa.

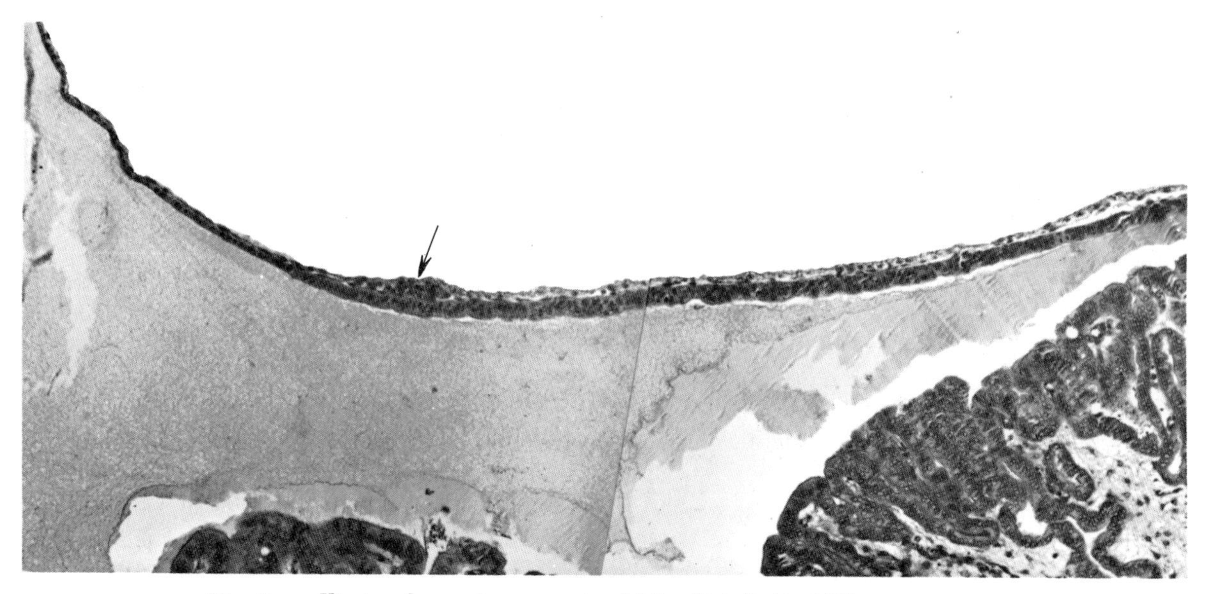

Fig. 2. — *Uterine horn of pregnant rabbit.* Detail (× 100).

OF GASTRULATION

Fig. 3. — *Detail of embryo* (× 300).

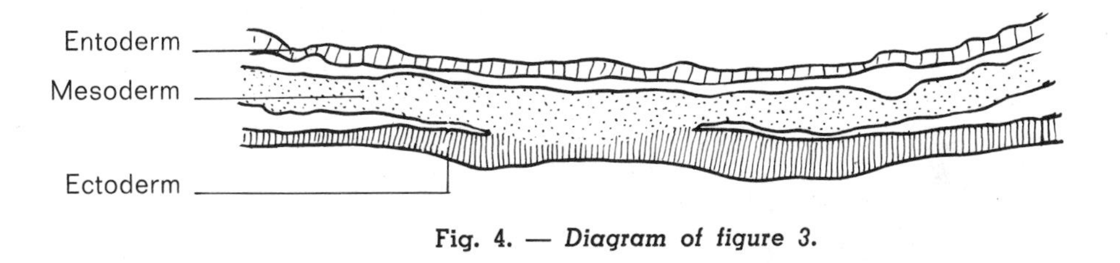

Fig. 4. — *Diagram of figure 3.*

Note that this dark line is far from being regular; in some places thickening or doubling of layers can produce the false impression that there is an embryo. Only examination at high magnification permits identification of an embryo. The primitive streak corresponds to the region where the ectoderm is fused and continuous with the mesoderm (part boxed in figure 2 and enlarged in figure 3).

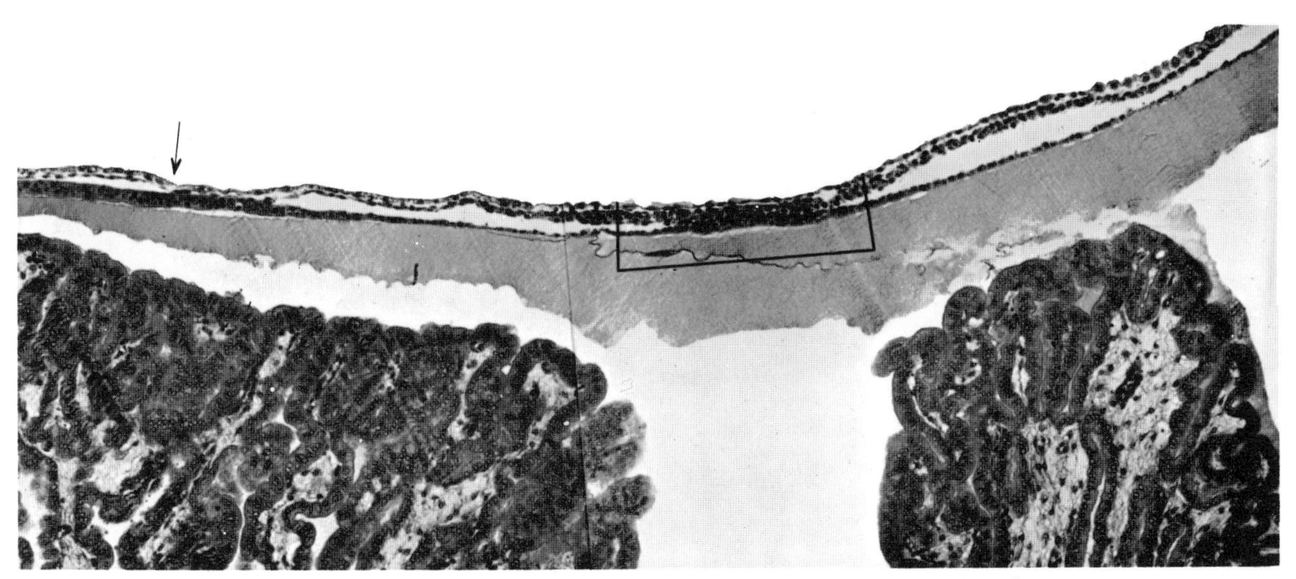

Fig. 2 *(continued).*

Passage from the prochordal plate stage to the definitive notochord stage begins at the cephalic end, then extends in the craniocaudal direction. Because of this, the appearance of cross sections of the embryo during gastrulation differs greatly according to the level examined.

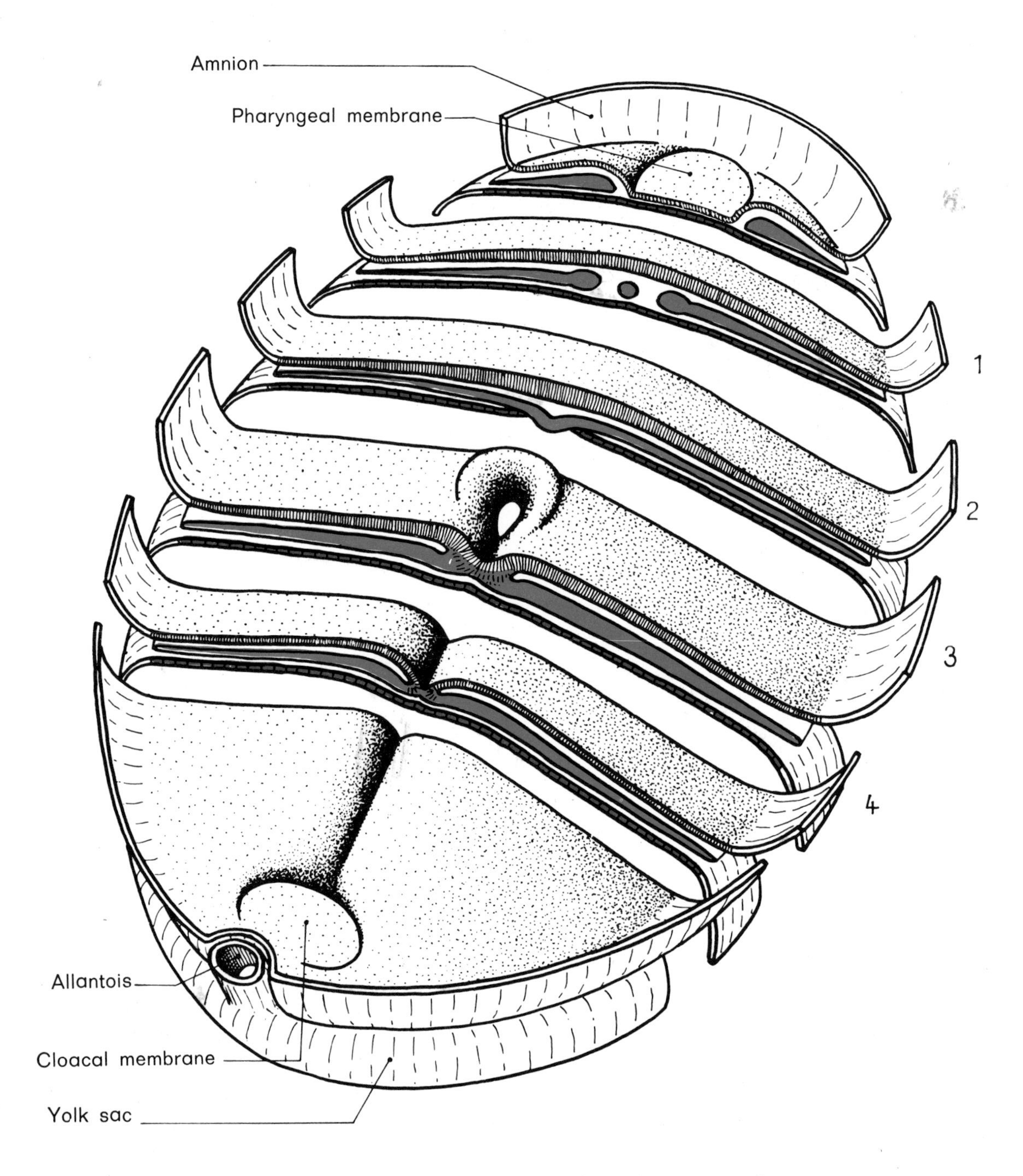

Diagrammatic view in cross section of embryo during gastrulation;
the numbers of the sections correspond to the photographs on the opposite page.

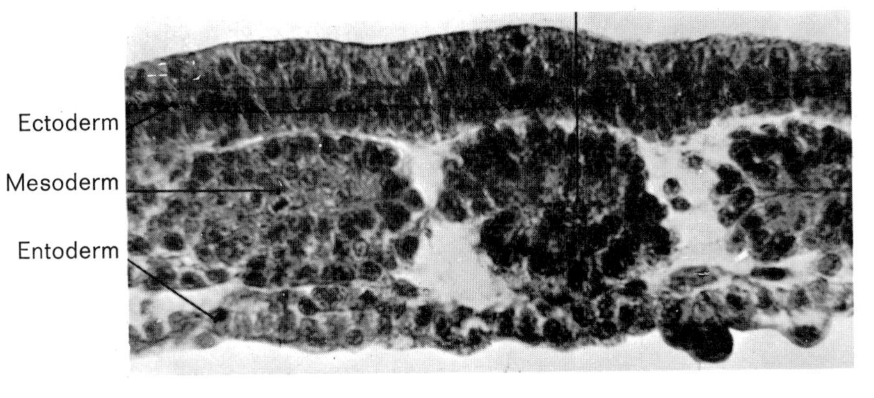

Fig. 1. — ***Section at level of the notochord.***

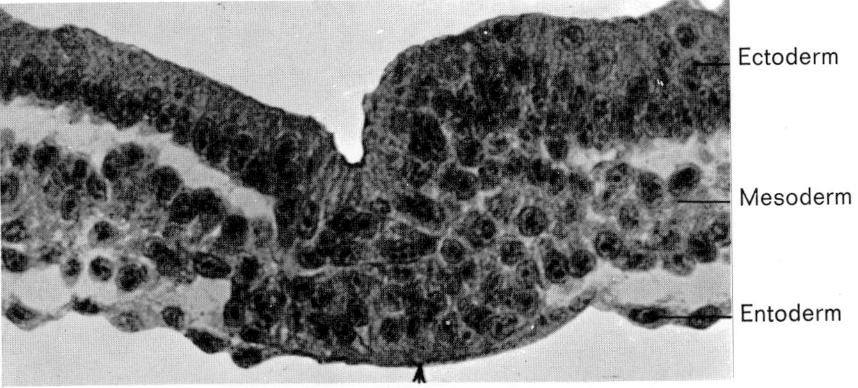

Fig. 2. — ***Section at level of the prochordal plate.***

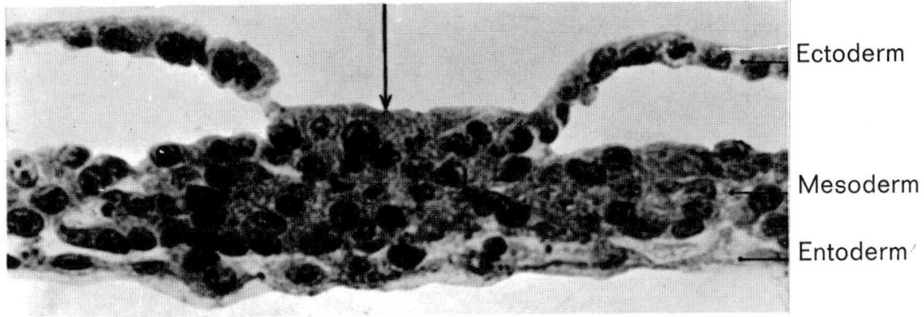

Fig. 3. — ***Section at level of Hensen's node.***

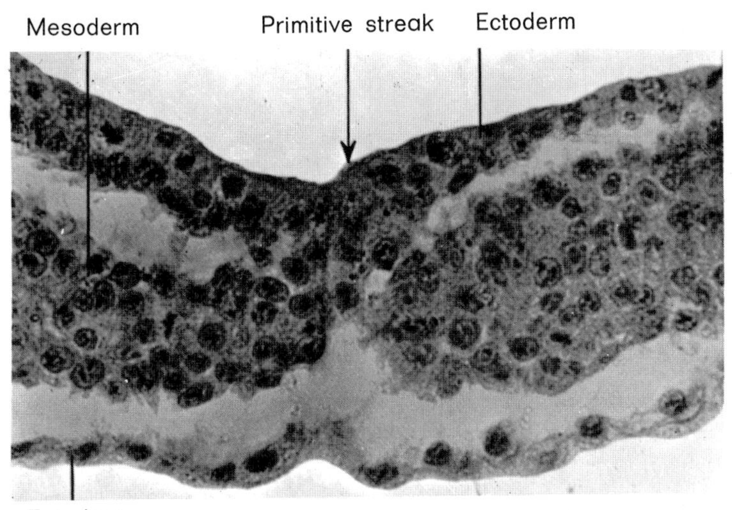

Fig. 4. — ***Section at level of primitive streak.***

Note. — These 4 sections are not from the same rabbit embryo. They were juxtaposed to illustrate the diagram opposite.

DESTINY OF

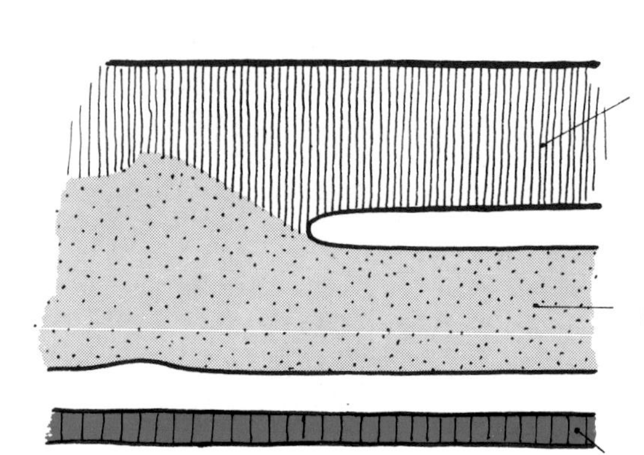

At the end of gastrulation, the three germ layers are in place.

In the following pages, we shall describe only the initial phases of development of each of these 3 layers.

THE THREE GERM LAYERS

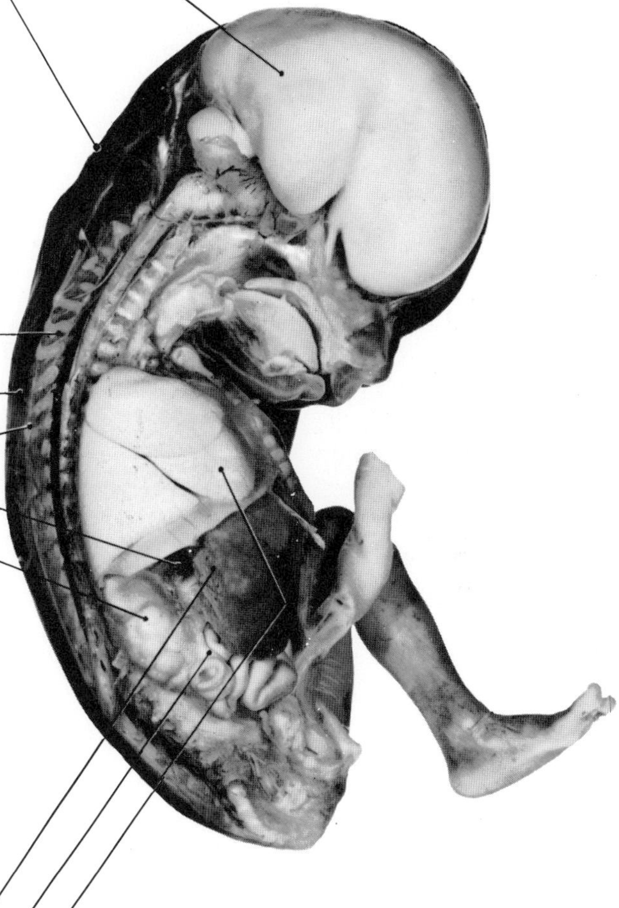

ECTODERM

NERVOUS TISSUE

EPIDERMIS

MESODERM

SKELETON

MUSCLE

CONNECTIVE TISSUE

CIRCULATORY SYSTEM

URINARY SYSTEM

ENTODERM

DIGESTIVE GLANDS

DIGESTIVE EPITHELIUM

RESPIRATORY EPITHELIUM

DEVELOPMENT OF THE ECTODERM :

The principal derivative of the ectoderm is nervous tissue or neural ectoderm. Its differentiation

I. — NEURAL PLATE AND NEURAL GROOVE STAGES

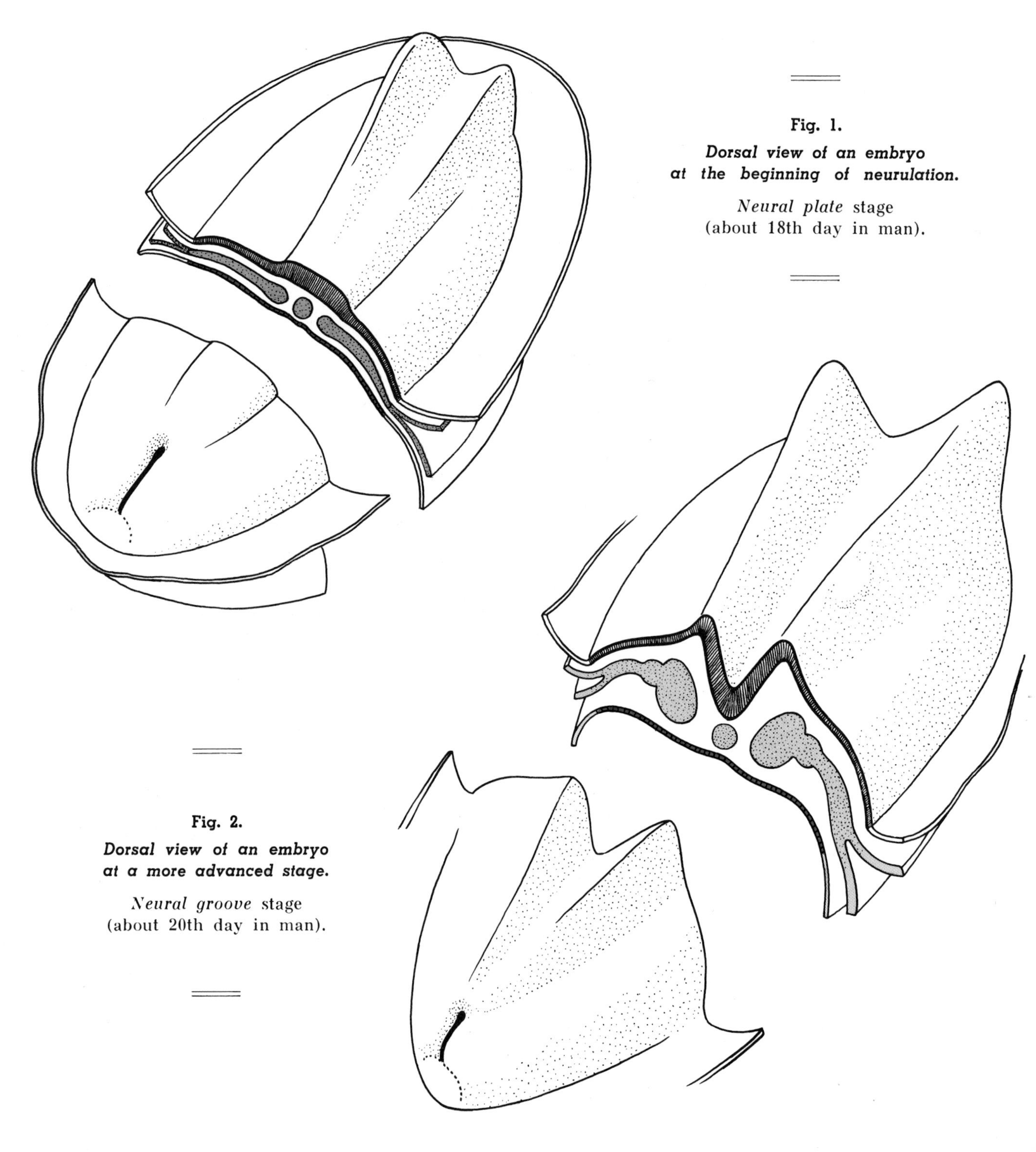

Fig. 1.
Dorsal view of an embryo at the beginning of neurulation.
Neural plate stage (about 18th day in man).

Fig. 2.
Dorsal view of an embryo at a more advanced stage.
Neural groove stage (about 20th day in man).

NEURULATION

constitutes ***neurulation.*** The rest of the ectoderm is then called ***surface ectoderm.***

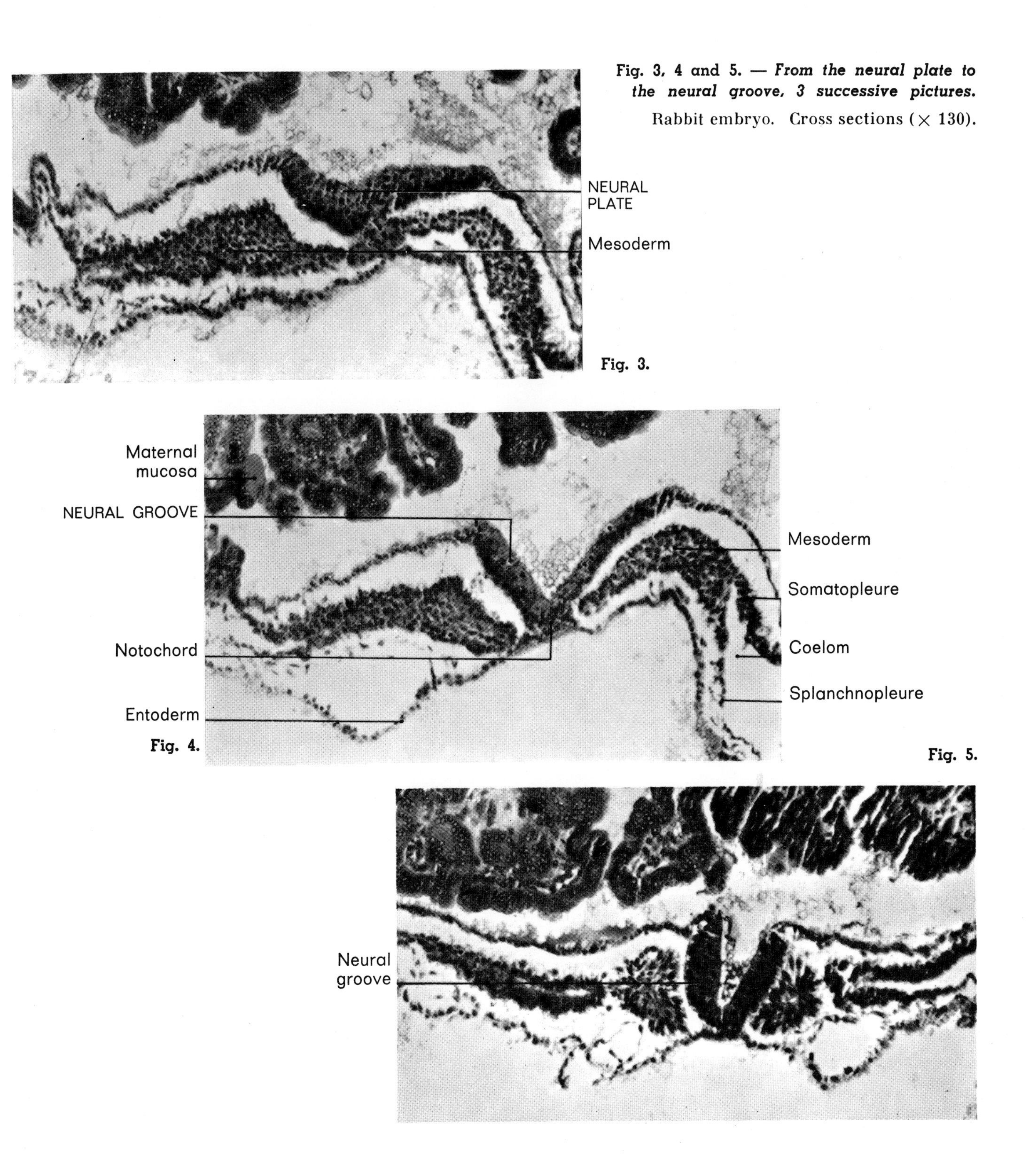

Fig. 3, 4 and 5. — *From the neural plate to the neural groove, 3 successive pictures.*
Rabbit embryo. Cross sections (× 130).

Fig. 3.

Fig. 4.

Fig. 5.

II. — NEURAL TUBE STAGE

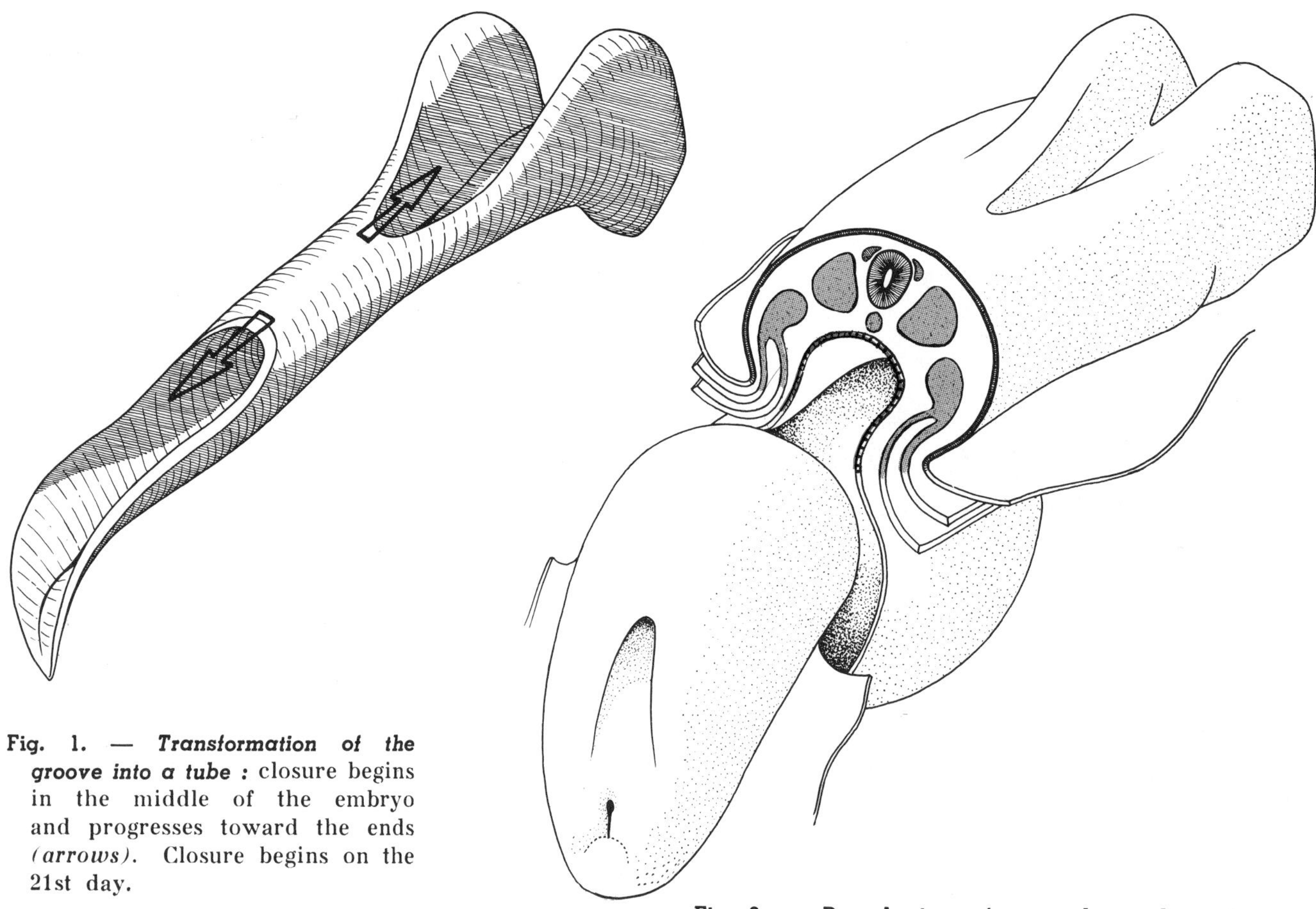

Fig. 1. — *Transformation of the groove into a tube :* closure begins in the middle of the embryo and progresses toward the ends *(arrows)*. Closure begins on the 21st day.

Fig. 2. — *Dorsal view of an embryo about the 22nd day :* the middle portion of the neural axis is already closed in tube form. The ends are still open as grooves.

At the time of tube closure, lateral cell clusters are detached, forming the neural crests easily visible in figure 3.

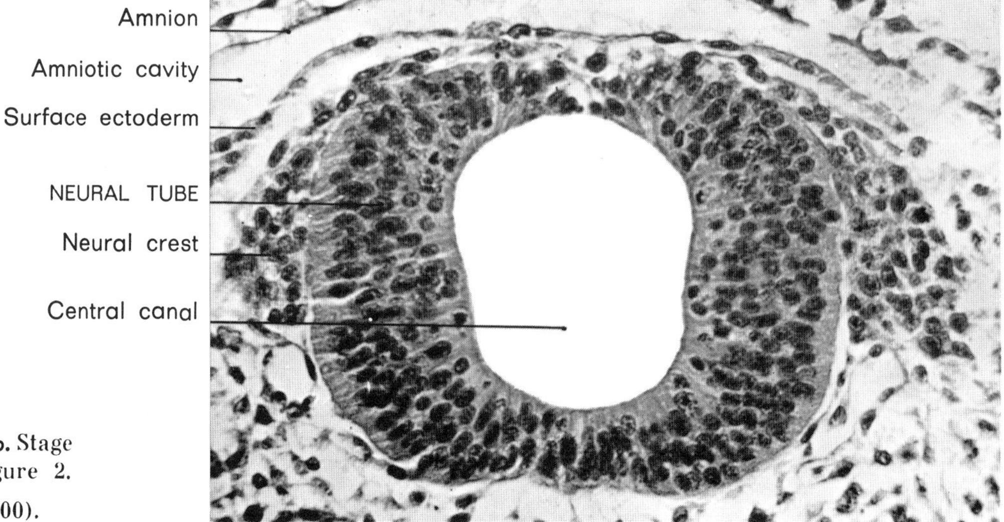

Fig. 3. – *Rabbit embryo.* Stage corresponding to figure 2. *Cross section* (× 300).

III. — CLOSURE OF ENDS OF THE TUBE

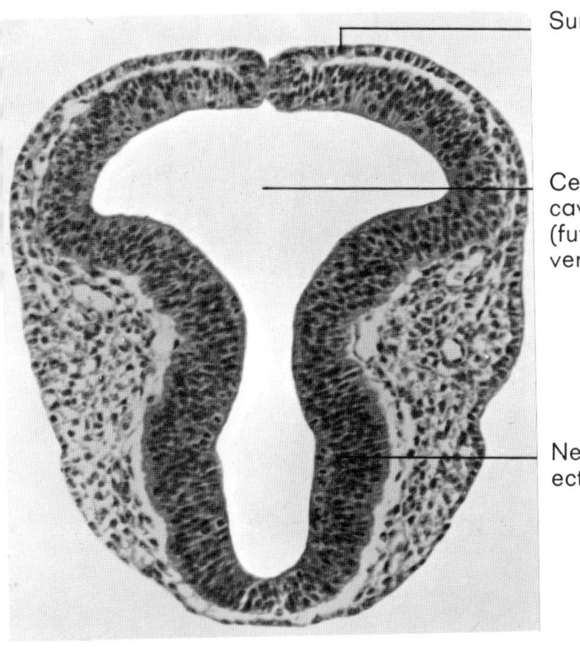

Fig. 4. — *Cross section of cephalic end of a rabbit embryo at time of closure of the anterior neuropore.*

Clearly visible in the upper part are the two edges of the neural groove touching each other on the median line. The neural groove has just closed.

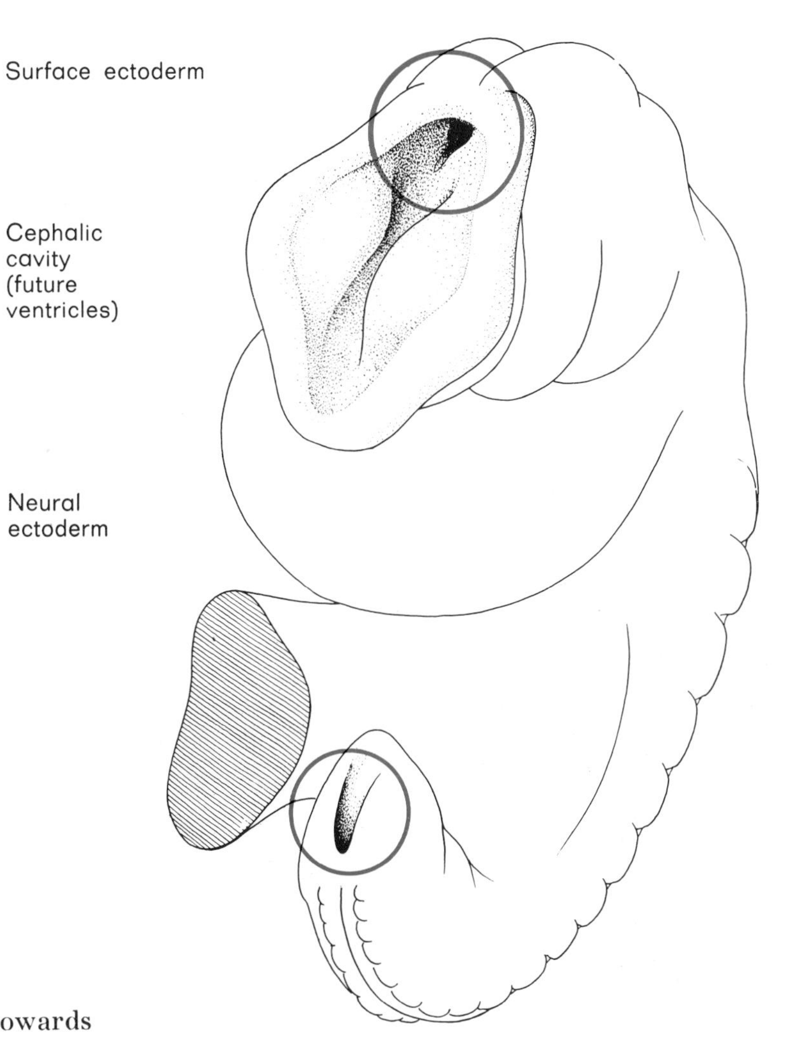

Fig. 5. — *Anterior and posterior neuropores of a 25th day human embryo.*

Progression of closure is more rapid towards the cephalic end than towards the caudal end.

The anterior or cranial neuropore closes towards the 26th day. The posterior or caudal neuropore closes towards the 28th day.

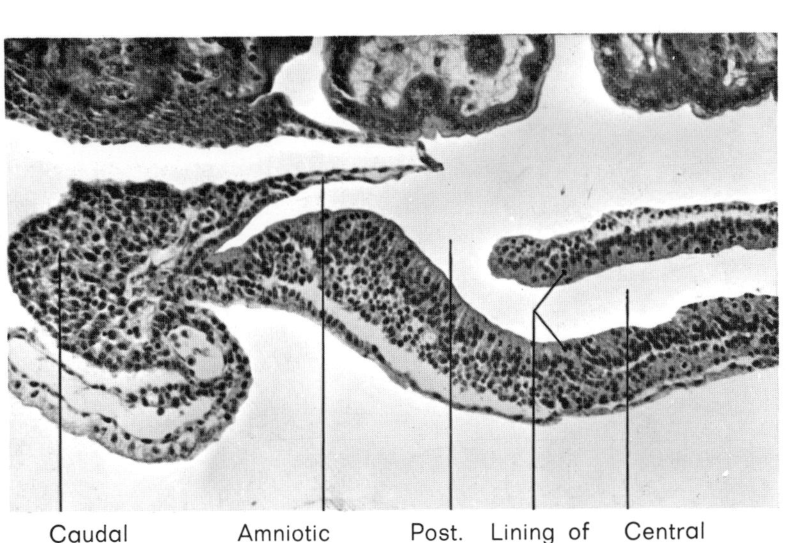

Fig. 6. — *Sagittal section of the caudal end of a rabbit embryo before closure of the posterior neuropore.* (The embryo is horizontal, its back upward toward the maternal mucosa.)

DEVELOPMENT

The development of the notochord, the paraxial mesoderm which gives rise to the somites, and the intermediate and lateral mesoderm will be studied successively.

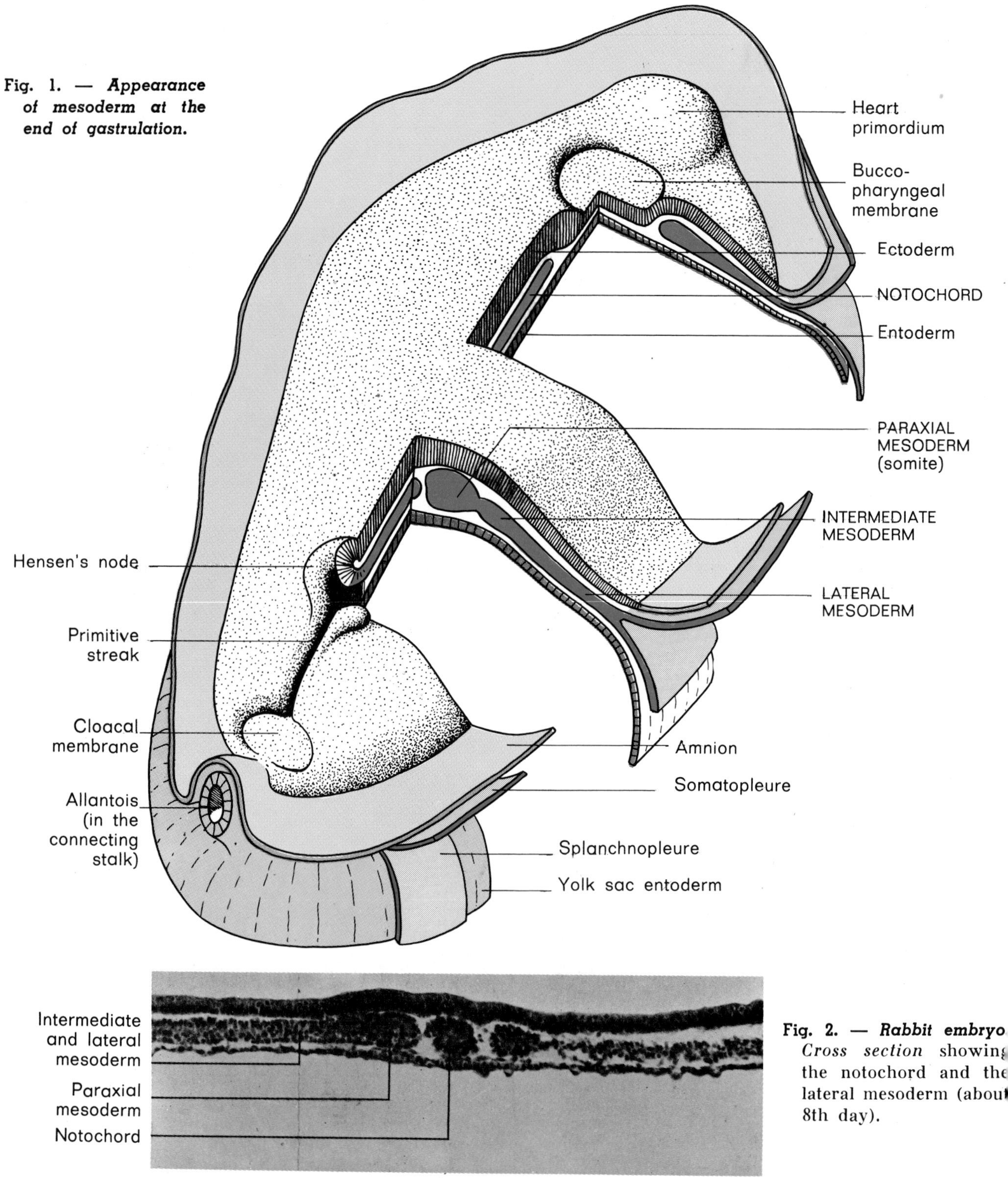

Fig. 1. — ***Appearance of mesoderm at the end of gastrulation.***

Fig. 2. — ***Rabbit embryo*** *Cross section* showin the notochord and the lateral mesoderm (abou 8th day).

OF THE MESODERM

I. — THE NOTOCHORD

The notochord forms the first longitudinal axis around which the vertebral bodies will be organized. Later, it regresses.

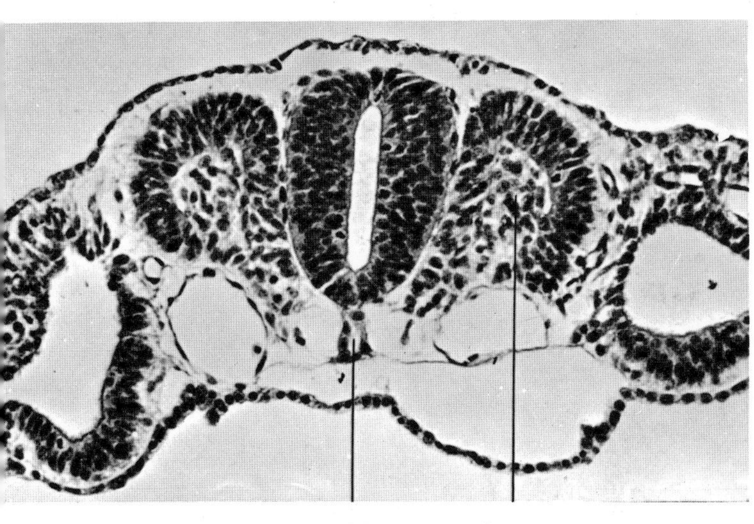

Fig. 3. — *Rabbit embryo.*
Cross section (about 10th day).

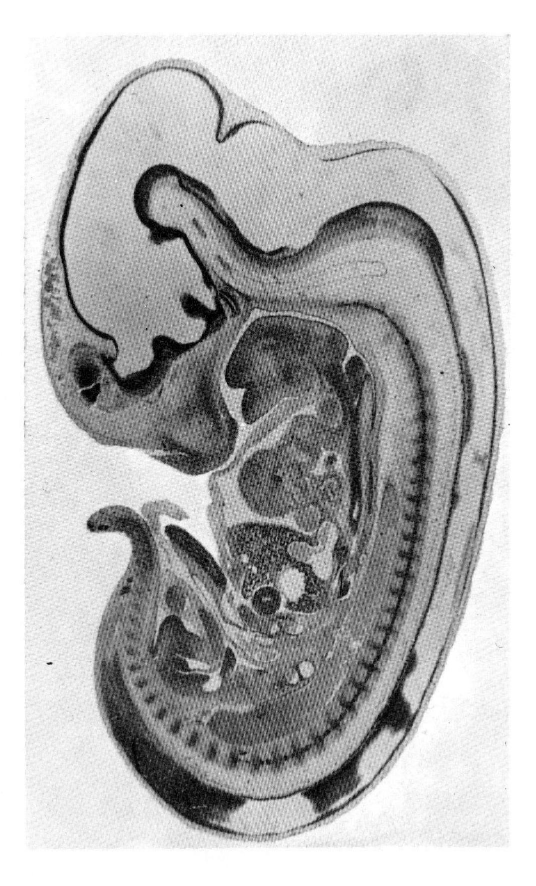

Fig. 4. — *Sagittal section of a rabbit embryo* (12-13th day). The notochord is visible throughout the length of the embryo. Around it, layered concentrations of cells represent the primordia of the future vertebral bodies.

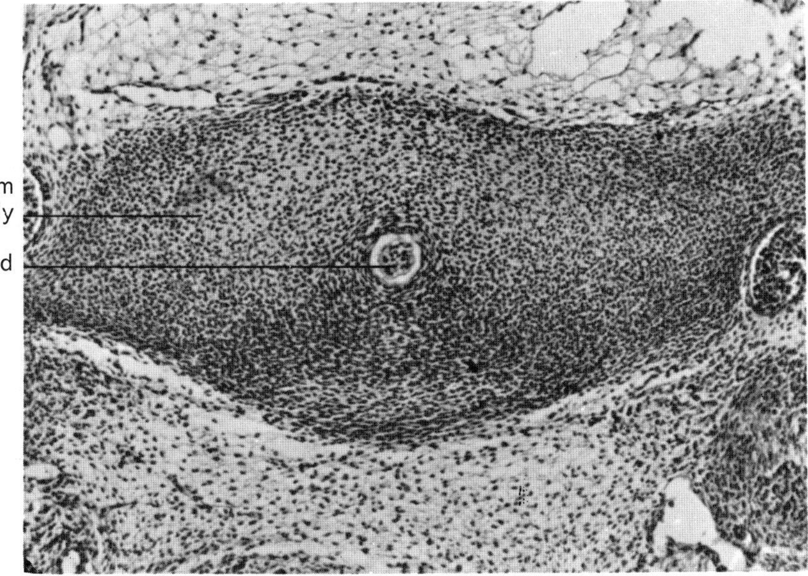

Fig. 5. — *Rabbit embryo* (12-13th day). Cross section of a vertebral primordium.

II. — THE SOMITES

Somites result from segmentation of the paraxial mesoderm.

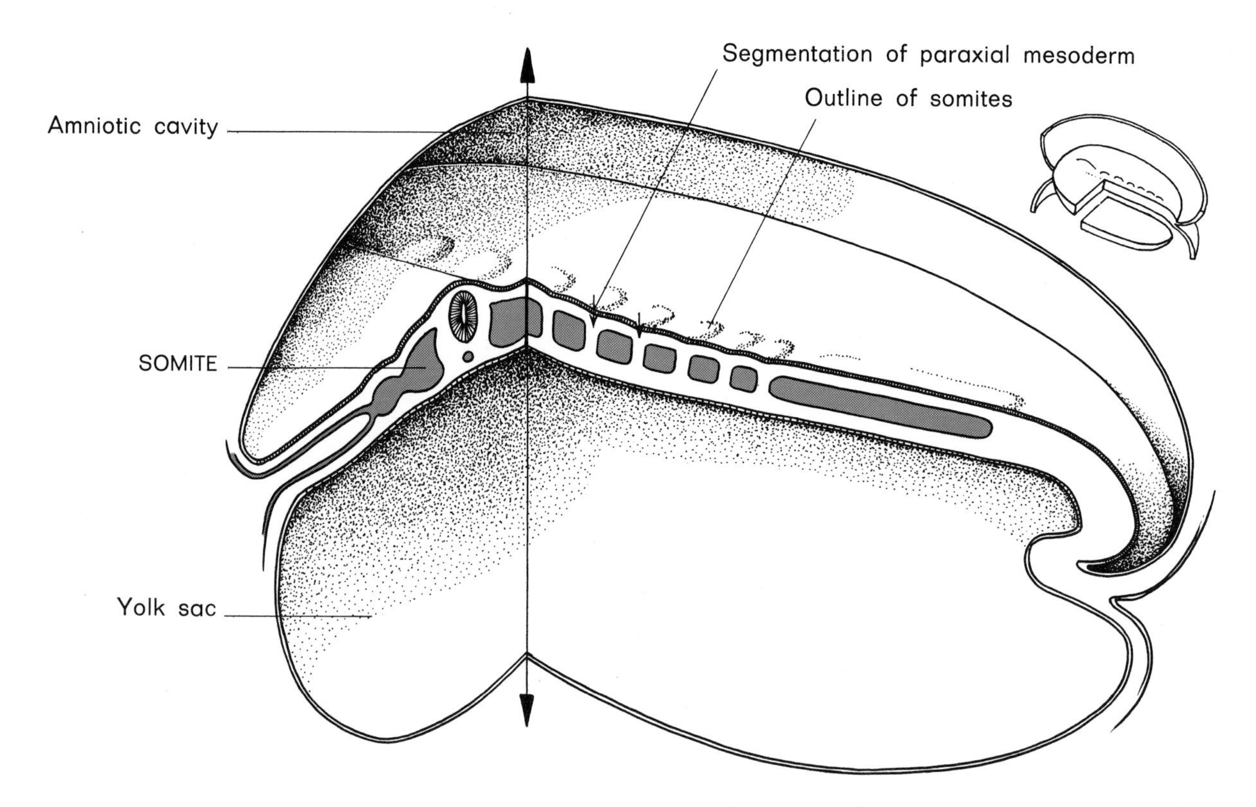

Fig. 1. — *Diagrammatic section of embryo along two perpendicular planes.* The arrow shows the angle of the two planes of section.

The paraxial mesoderm thickens and fragments metamerically.

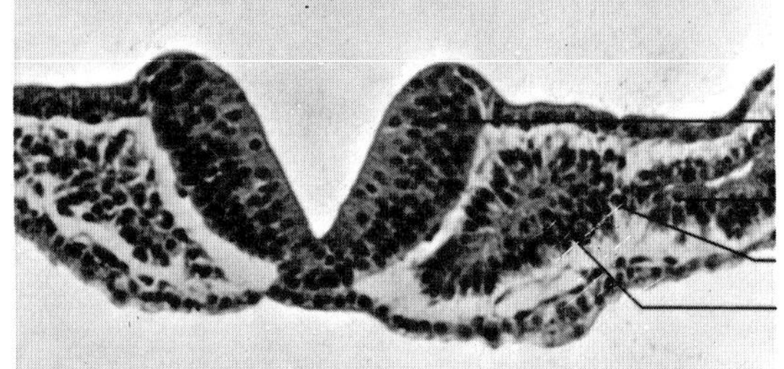

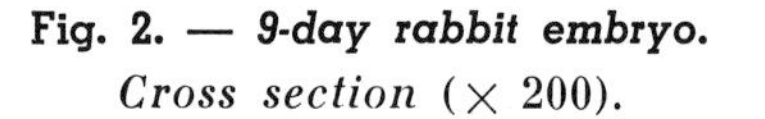

Fig. 2. — *9-day rabbit embryo.*
Cross section (× 200).

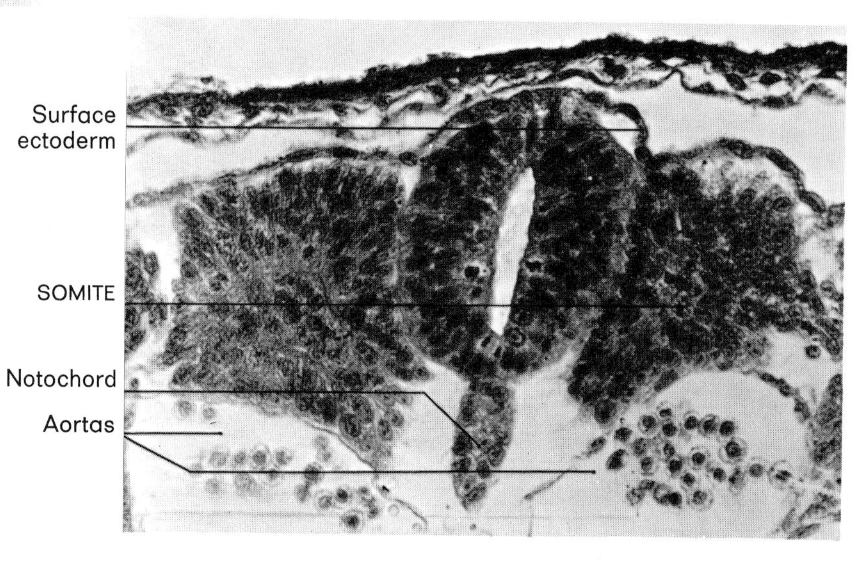

Fig. 3. — *10-day rabbit embryo.*
Cross section (× 300).

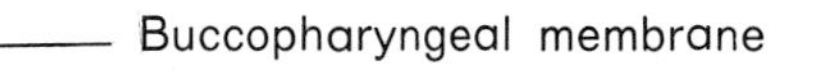

The first somites appear in the middle portion of the embryo. However, because of the predominant cephalic development following this stage, the region where the first somites appear actually corresponds to the future occipital area.

After appearance of the first somites, segmentation progresses toward the caudal region.

At the end of the 5th week, the human embryo has about 42 pairs of somites.

Fig. 4. — *Dorsal view of a human embryo of about 3 weeks.*

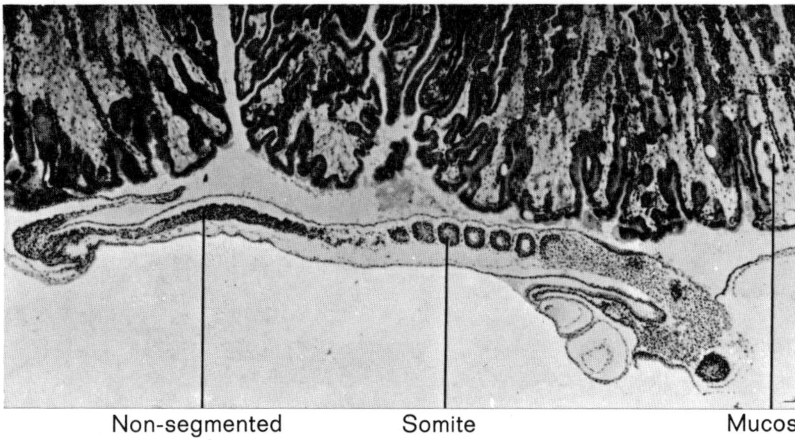

Fig. 5. — *Rabbit embryo (about 10th day).* *Longitudinal paramedian section* (× 30).

Seven well differentiated somites can be distinguished in the cephalic region. Towards the caudal end, the mesoderm is still spread out in an undivided sheet.

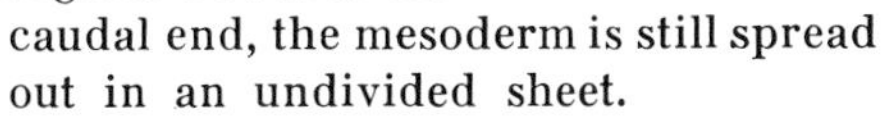

The ***sclerotome,*** which gives rise to the vertebral primordia, and the ***myotome,*** which gives rise to part of the muscle system, both originate from the ***somites.***

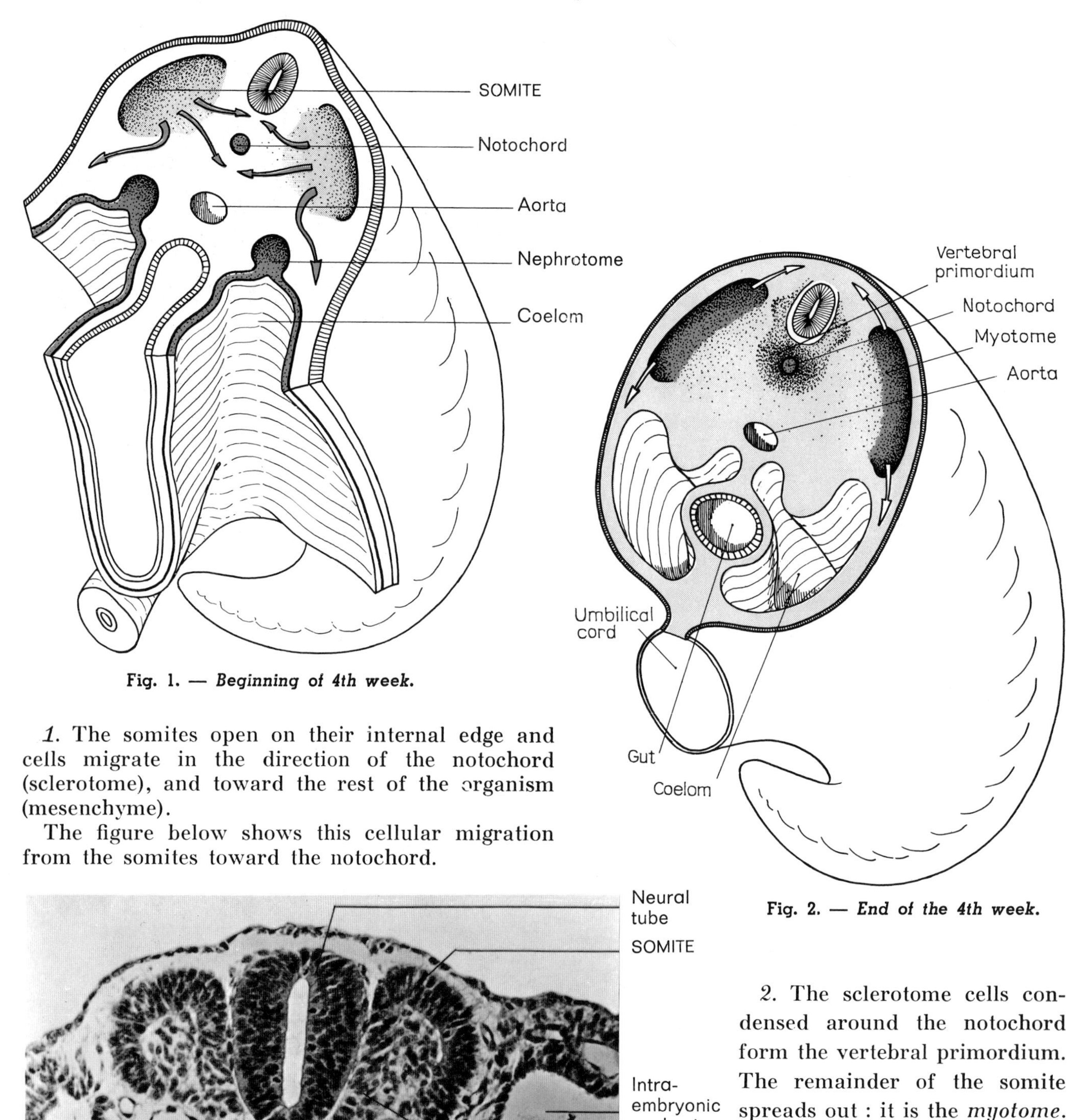

Fig. 1. — *Beginning of 4th week.*

Fig. 2. — *End of the 4th week.*

1. The somites open on their internal edge and cells migrate in the direction of the notochord (sclerotome), and toward the rest of the organism (mesenchyme).

The figure below shows this cellular migration from the somites toward the notochord.

2. The sclerotome cells condensed around the notochord form the vertebral primordium. The remainder of the somite spreads out : it is the *myotome.*

Fig. 3. — *Rabbit embryo, 9-10 days.*
Cross section (× 150).

NOTE

This transformation of the somites is illustrated in 4 diagrams drawn to the same scale. Actually, between the first and the last stages pictured, the embryo grows considerably in volume.

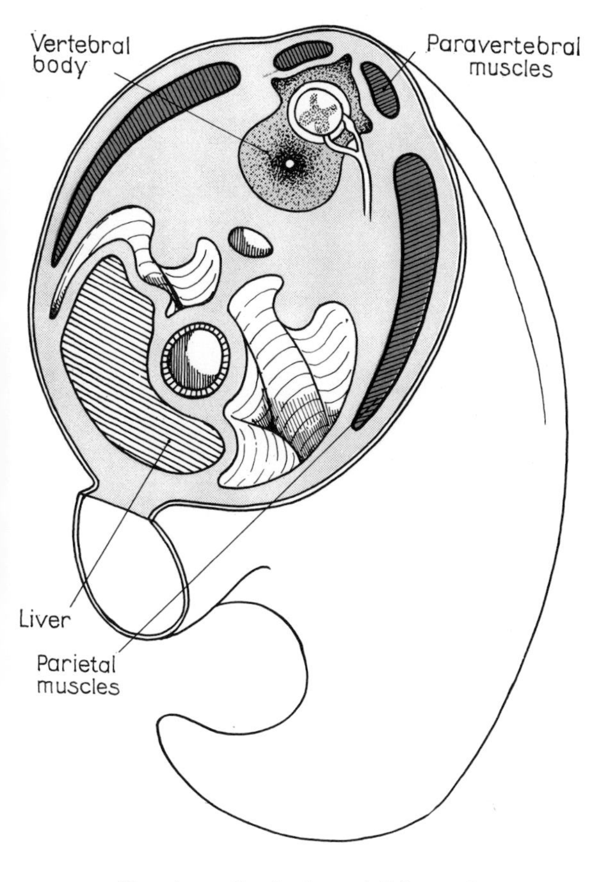

Fig. 4. — *Beginning of 5th week.*

3. The myotome gives rise to the vertebral muscles...

4. ... and, with the somatopleure, those of the limbs and the anterior lateral body wall.

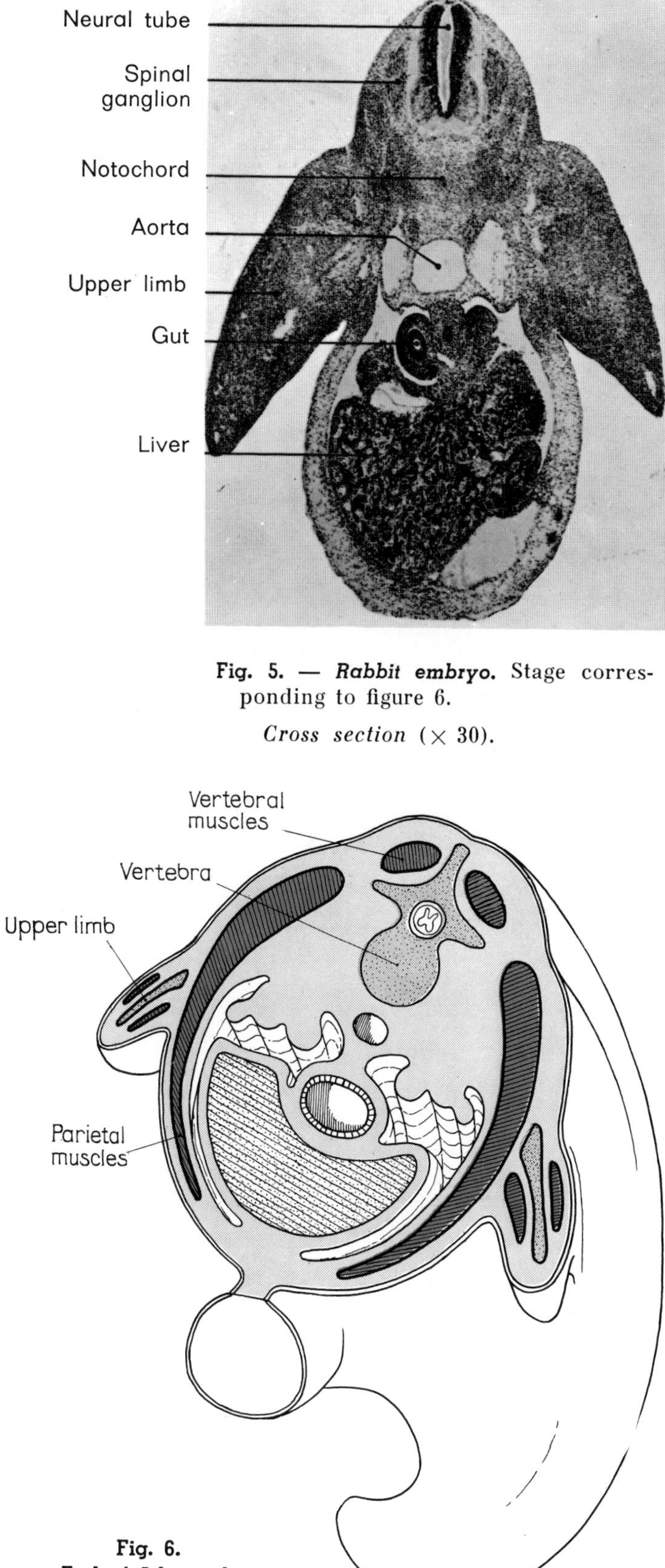

Fig. 5. — *Rabbit embryo.* Stage corresponding to figure 6.
Cross section (× 30).

Fig. 6.
End of 5th week.

III. — THE INTERMEDIATE

The ***nephrogenic cord,*** which gives rise to the excretory apparatus, originates from the *intermediate mesoderm.* This aggregation of mesodermal cells undergoes a metameric segmentation parallel to that of the somites and forms the *nephrotomes.* However, the segmentation is never completed, and does not reach the caudal part of the nephrogenic cord.

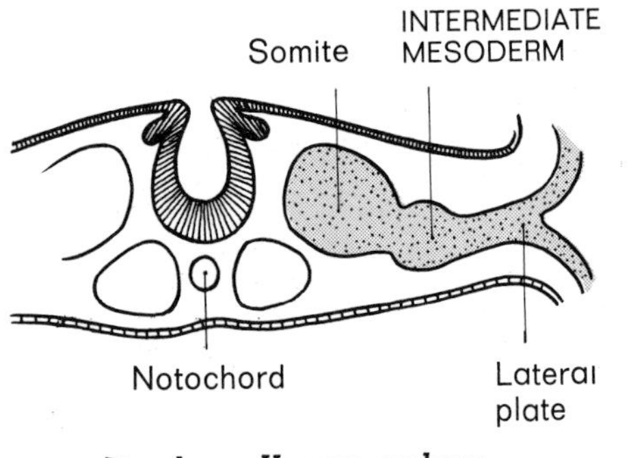

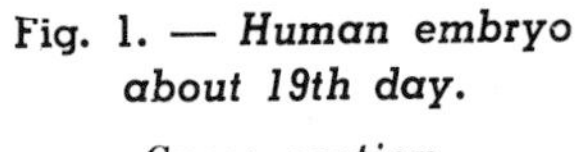

Fig. 1. — *Human embryo about 19th day.*
Cross section.

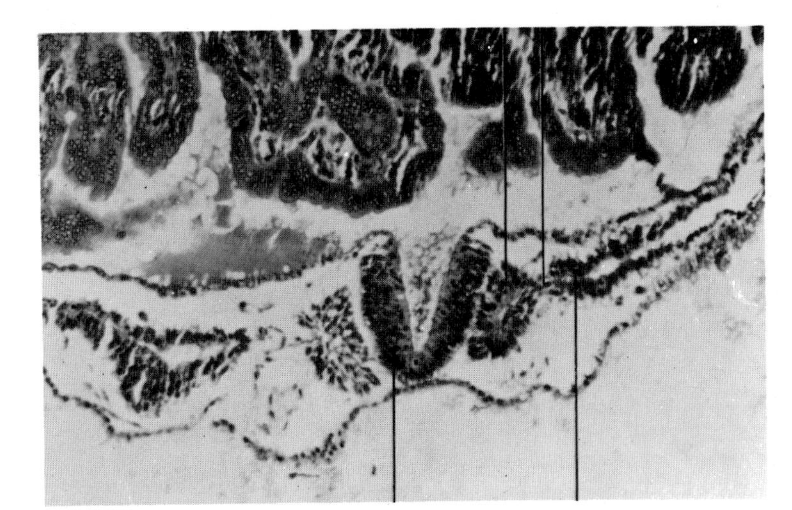

Fig. 2. — *Rabbit embryo at a stage corresponding to figure 1.*
Cross section.

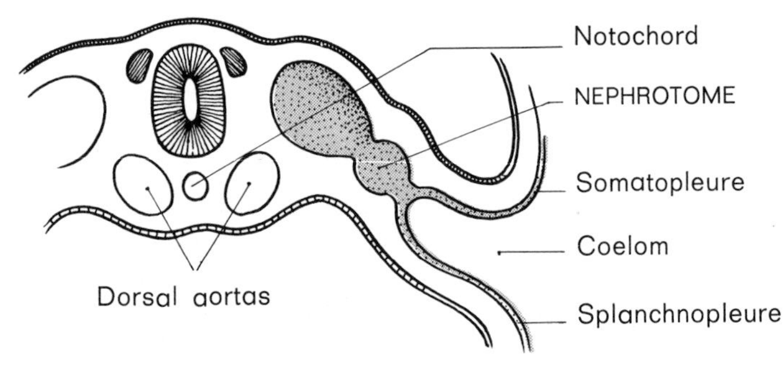

Fig. 3. — *Human embryo about 21st day.*
Cross section.

Fig. 4. — *Rabbit embryo at a stage corresponding to figure 3.*
Cross section.

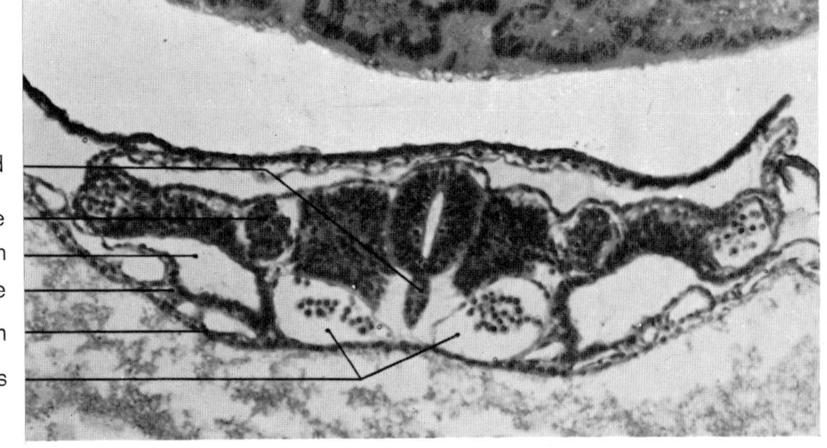

AND LATERAL PLATE MESODERM

The lateral plate splits into two layers :

— One layer lines the entoderm, and continues in the extraembryonic splanchnopleure covering the yolk sac. This is the *intraembryonic splanchnopleure* which gives rise to the muscle and connective tissue layers of the trunk.

— The other layer of lateral plate cells lines the ectoderm and continues in the *intraembryonic somatopleure* which participates in formation of the lateral and ventral trunk walls.

Between the two layers of the lateral plate a cavity prolonging the extra-embryonic coelom appears : it is the intraembryonic coelom, future pleural-pericardial-peritoneal cavity.

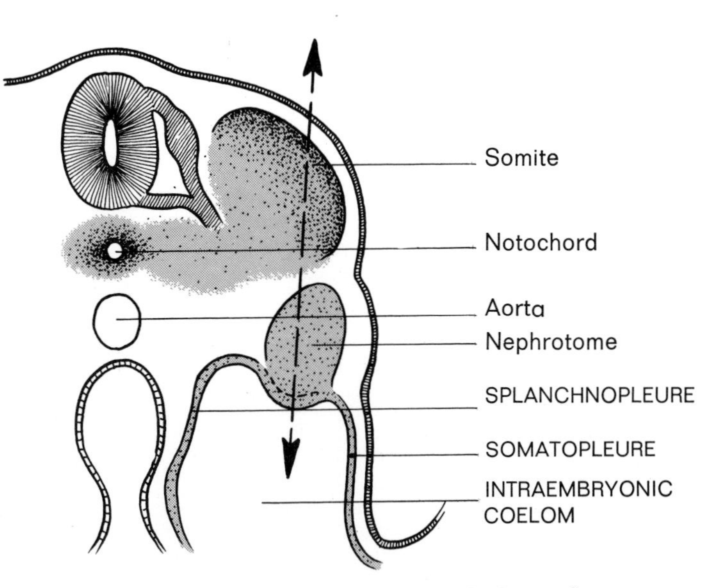

Fig. 5. — *Human embryo about 26th day.* *Cross section.* The arrow represents the plane of section of figure 6.

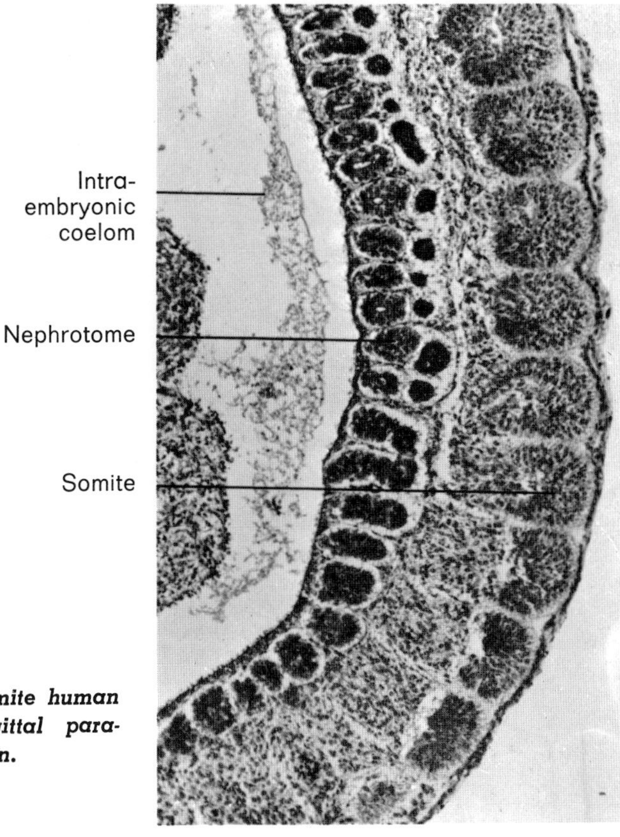

Fig. 6. — *30 somite human embryo. Sagittal para-median section.*

DEVELOPMENT

Development of the entoderm is simpler than that of the other two germ layers : until cephalocaudal flexion* occurs, it is a monocellular layer forming the lining of the yolk sac.

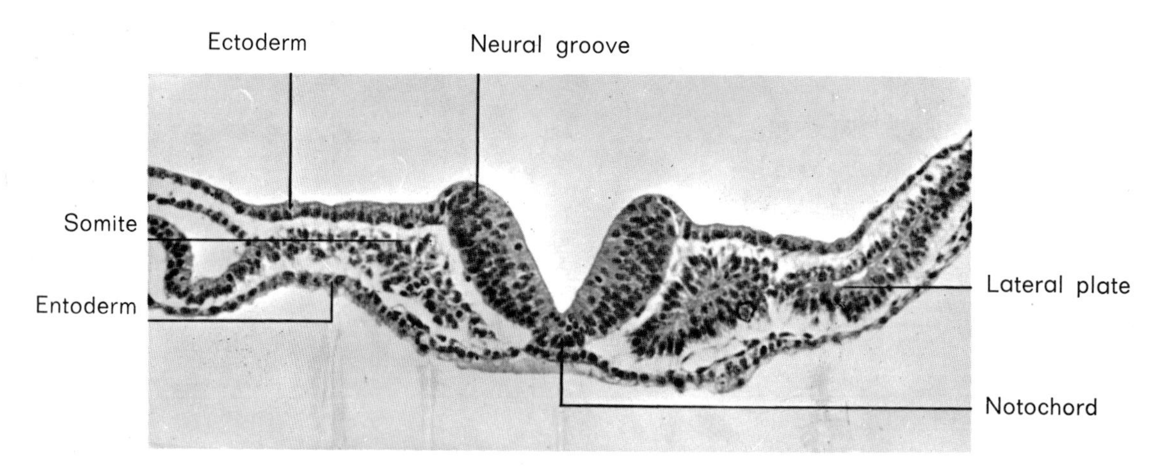

Fig. 1. — *8-day rabbit embryo.* *Cross section.*
Entoderm immediately after gastrulation.

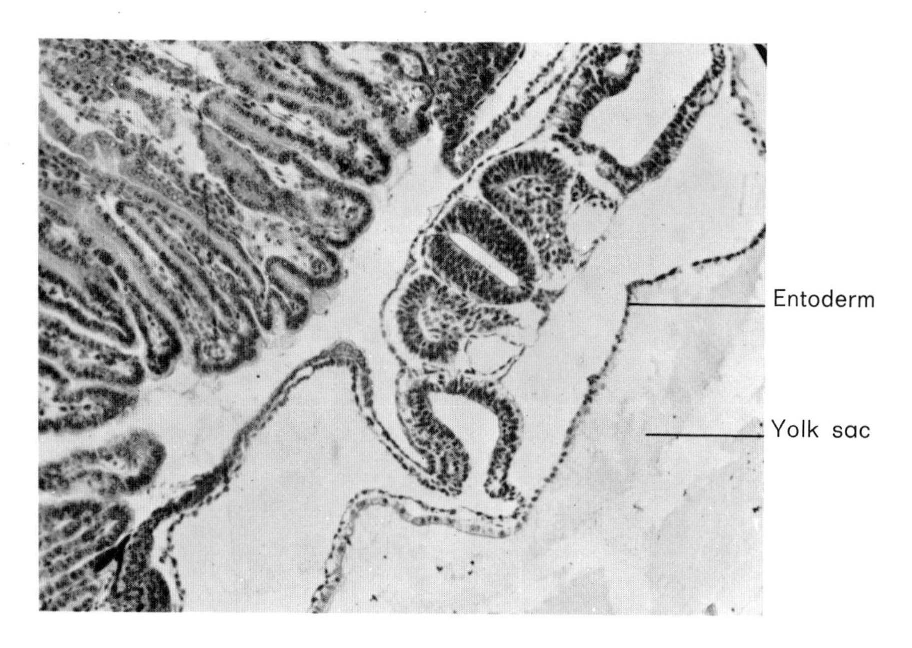

Fig. 2. — *9-day rabbit embryo.* *Cross section.*
Entoderm at the end of neurulation.

* See footnote on p. 46.

OF THE ENTODERM

The primitive gut originates from the entoderm at the time of cephalocaudal flexion of the embryo.

I. — FORMATION OF PRIMITIVE GUT

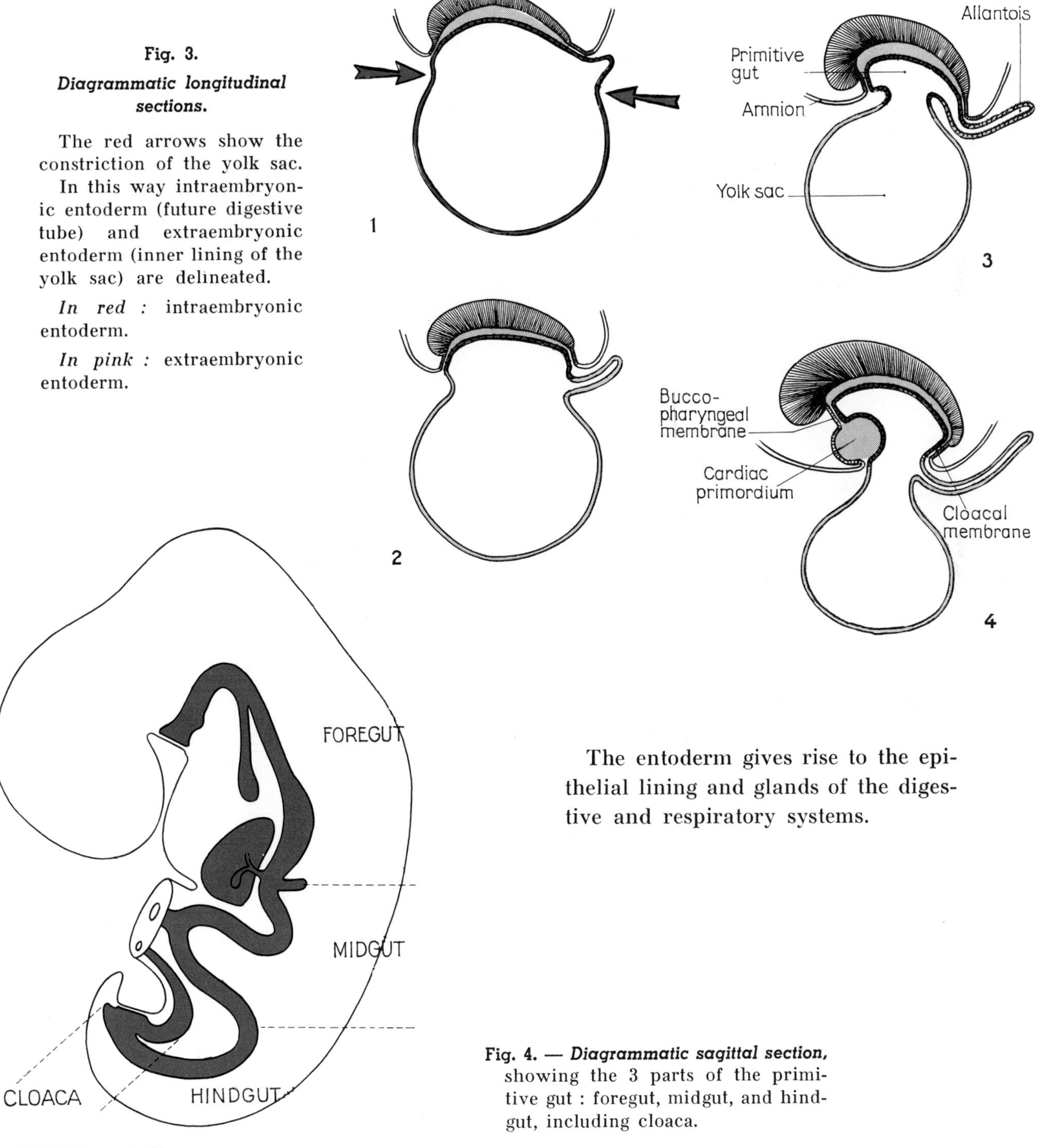

Fig. 3.
Diagrammatic longitudinal sections.

The red arrows show the constriction of the yolk sac. In this way intraembryonic entoderm (future digestive tube) and extraembryonic entoderm (inner lining of the yolk sac) are delineated.

In red : intraembryonic entoderm.

In pink : extraembryonic entoderm.

The entoderm gives rise to the epithelial lining and glands of the digestive and respiratory systems.

Fig. 4. — ***Diagrammatic sagittal section,*** showing the 3 parts of the primitive gut : foregut, midgut, and hindgut, including cloaca.

II. — THE BUCCOPHARYNGEAL AND CLOACAL MEMBRANES

The buccopharyngeal and cloacal membranes temporarily close the two ends of the primitive gut. These two regions where ectoderm and entoderm are in direct contact, without interposition of mesoderm, later disappear.

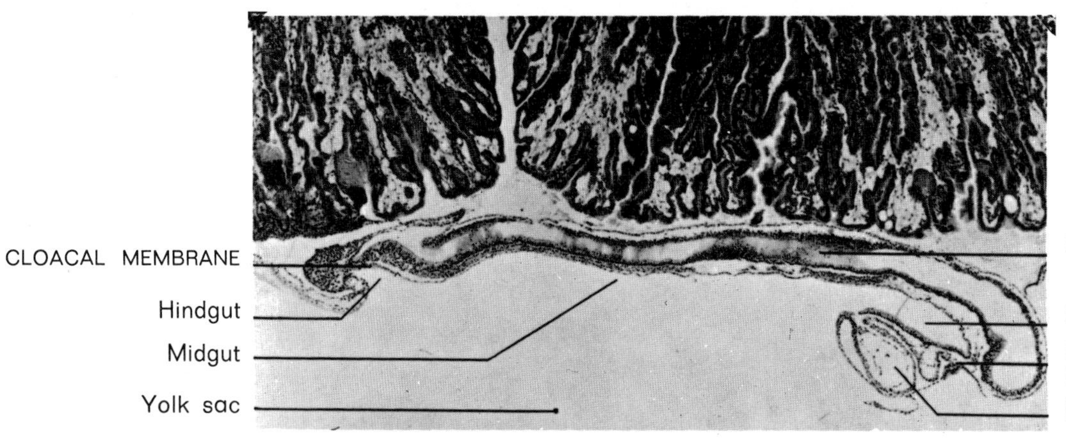

Fig. 1. — *Rabbit embryo.* *Median sagittal section* (× 30).

THE BUCCOPHARYNGEAL MEMBRANE

In man, the buccopharyngeal membrane disappears at the beginning of the 4th week.

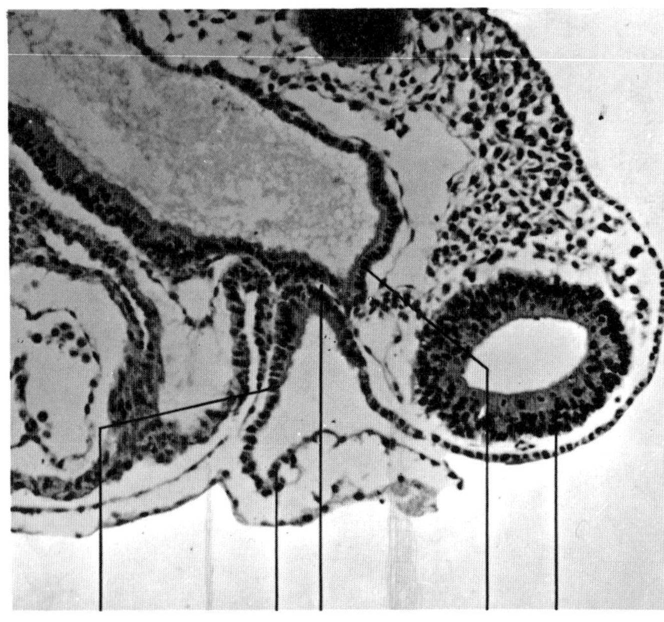

Fig. 2. — *Rabbit embryo.*
Sagittal section (× 140).

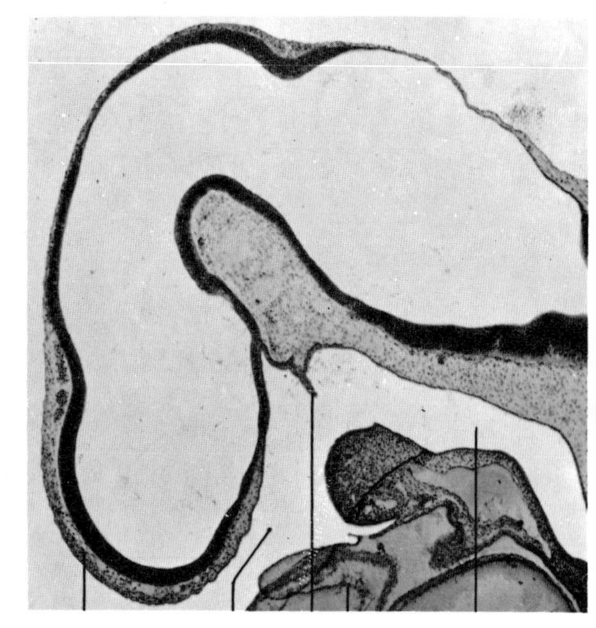

Fig. 3. — *Rabbit embryo.*
Sagittal section (× 40).

* BM : remnant of buccopharyngeal membrane.

THE CLOACAL MEMBRANE

This membrane persists longer than the buccopharyngeal membrane. When the embryo is about 7 weeks old, the cloacal membrane, like the cloaca, is divided into two parts :—

— a posterior, *anal membrane;* — an anterior, *urogenital membrane.*

The anal membrane is resorbed about the 9th week.

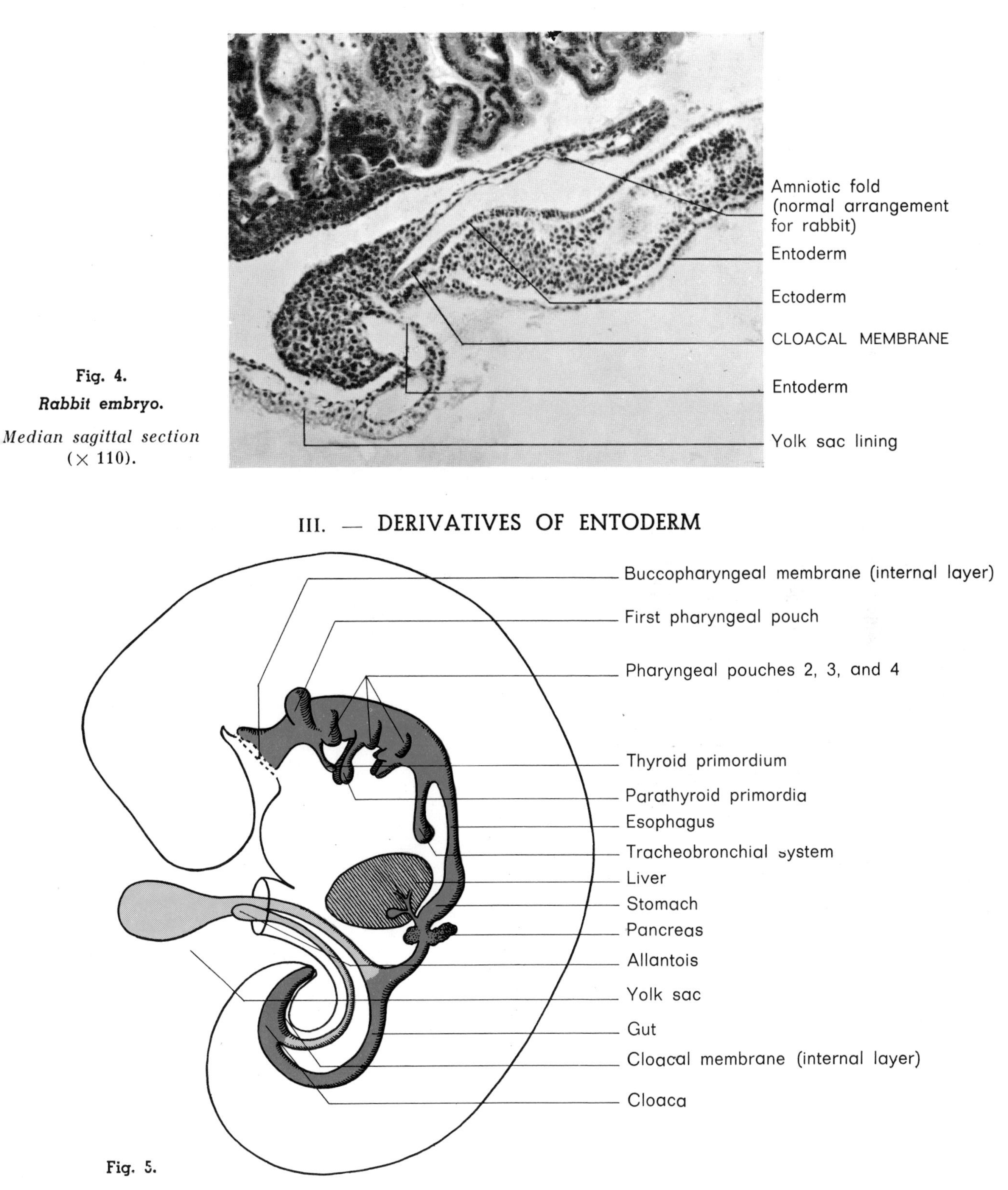

Fig. 4.
Rabbit embryo.
Median sagittal section (× 110).

III. — DERIVATIVES OF ENTODERM

Fig. 5.

FLEXION *

Until gastrulation the embryo is a disc continuous with its accessory tissues all around its circumference.

The phenomenon of flexion, a process of curving, transforms the embryo into a tube and isolates it from the extraembryonic membranes, to which it is eventually joined only by a thin stalk, the umbilical cord.

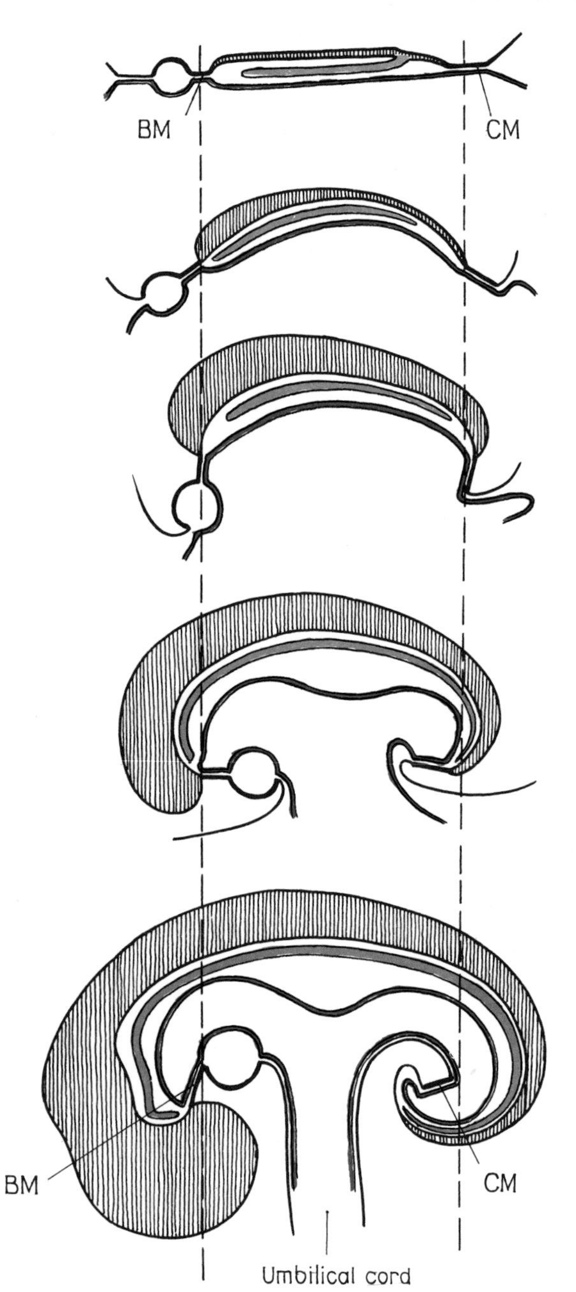

Fig. 1. — *Sequence of flexion in longitudinal section.* The dorsal region grows very rapidly. The ventral region varies little. The embryo curves itself around the umbilical region. BM = buccopharyngeal membrane. CM = cloacal membrane.

Fig. 2. — *Sequence of flexion in cross section.* The dorsal region thickens, especially on the midline. The edges of the disc swing ventrally carrying the amnion with them. Finally, the embryo is surrounded by the amniotic cavity.

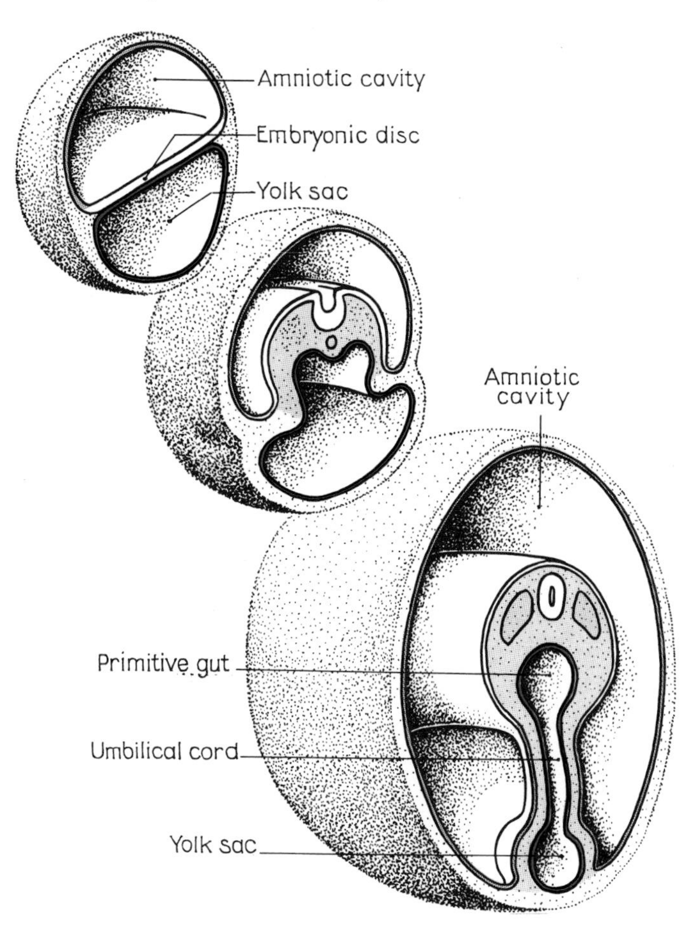

* *Translator's note :* French embryologists use the term *delimitation* to denote the overall process involving formation of the body cylinder and cephalocaudal flexion, and leading from the flat embryonic disc to assumption of the basic embryonic body form. The terms cephalocaudal flexion, or flexion, will be used here.

Flexion will be studied successively :

— *in cross section* (pp. 48 to 51), by sections through the different levels shown in figures 3 and 4 on this page.

— *longitudinally* (pp. 52 and 53), by a median section.

Fig. 3. — *The diagram opposite gives an overall view of the embryo and surrounding structures shown in the subsequent figures.* The surrounding membranes are partially removed to show the embryo.

The black arrow indicates the plane of section of the diagram below.

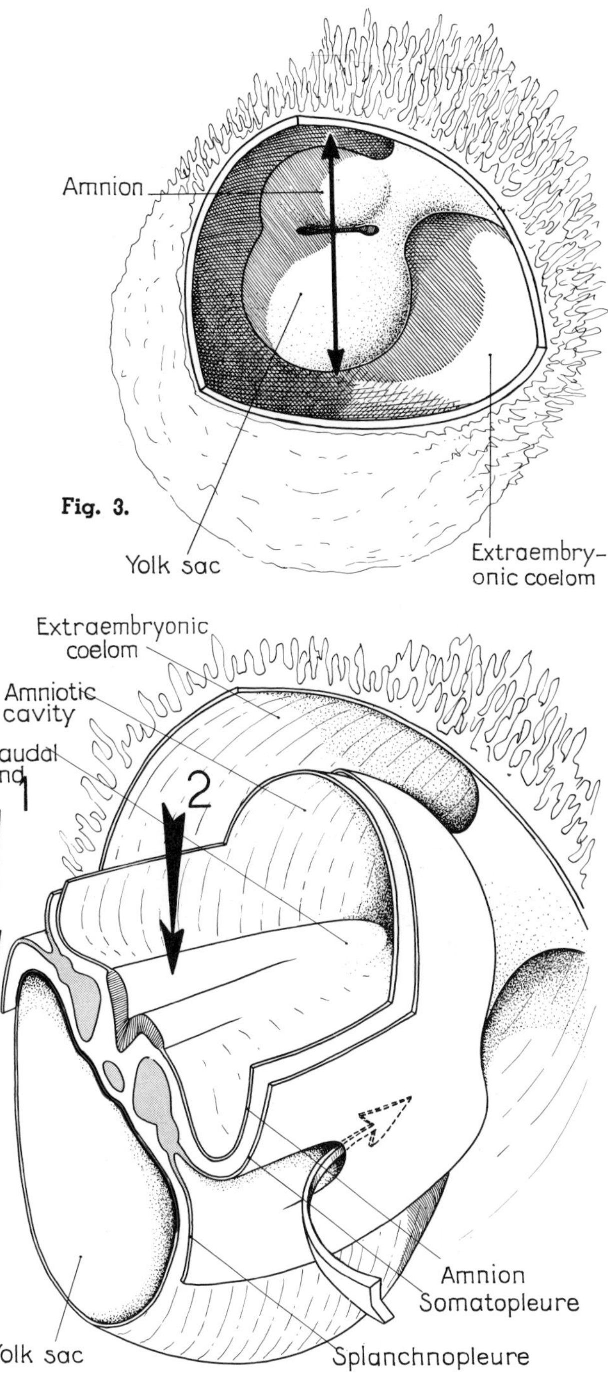

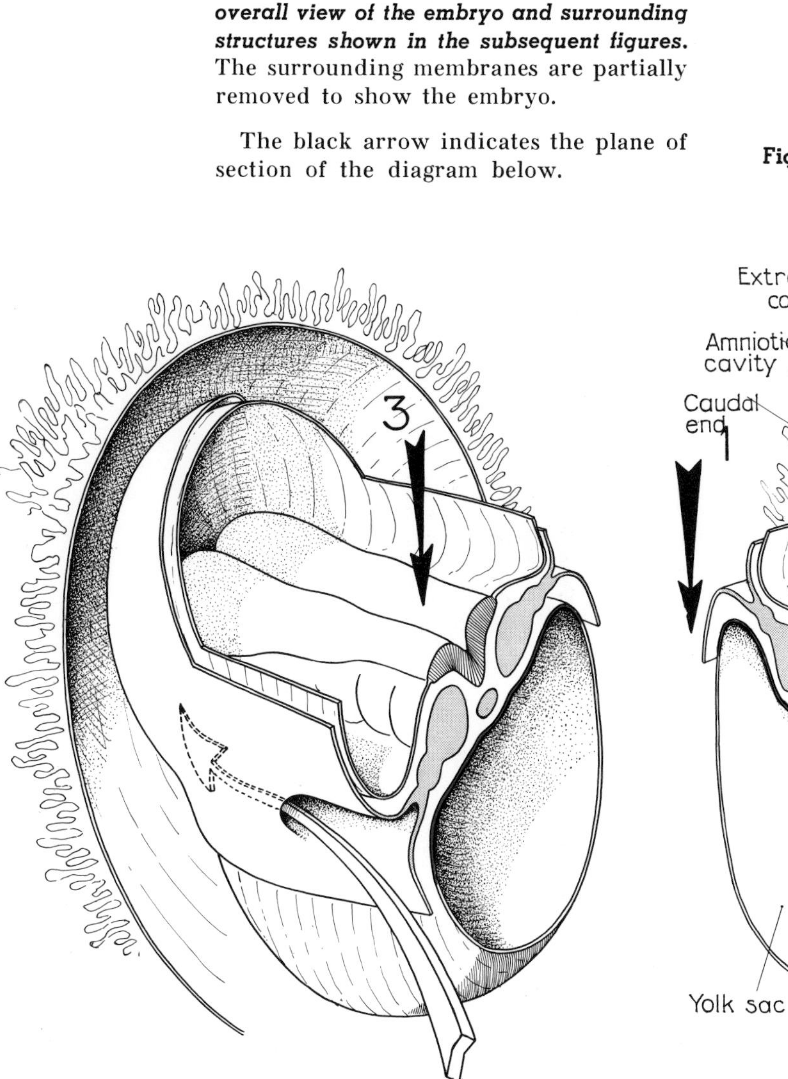

Fig. 4. — *The amniotic cavity was opened and the embryo was cross sectioned* (arrow 1). Arrows 2 and 3 show other planes of section, studied in the following pages.

— Plane 1 passes through the umbilical region.
— Plane 2 is caudal to the umbilical region.
— Plane 3 is cephalic to the umbilical region.

After this stage, the extraembryonic coelom has 2 intraembryonic prolongations, shown by the two white arrows. As a result of the movements involved in flexion, these 2 prolongations will form the cranial part and the caudal part of the intraembryonic coelom.

FLEXION

I. — UMBILICAL REGION

In the umbilical region, closure remains incomplete until birth; here the structures of the umbilical cord pass into the embryo.

a) Before the beginning of flexion (fig. 1), the embryo is flat, between the amniotic cavity and the yolk sac.

b) The lateral edges swing ventrally (fig. 2, arrows). The yolk sac begins to differentiate.

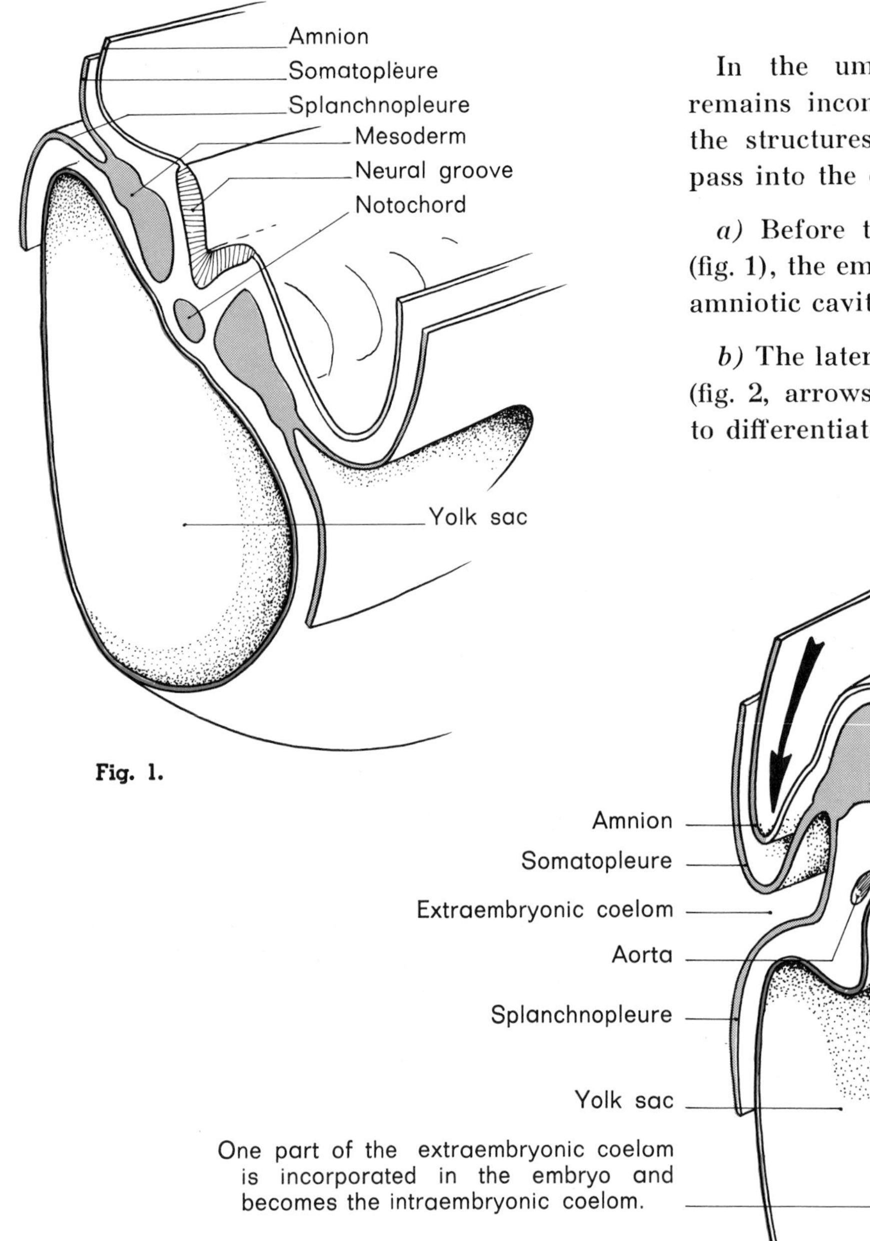

Fig. 1.

Fig. 2.

IN CROSS SECTION

c) The ventral wall begins to form (fig. 3). The gut is delineated, surrounded by intra-embryonic coelom.

d) At the level of the umbilicus, the ventral wall does not close completely (fig. 4).

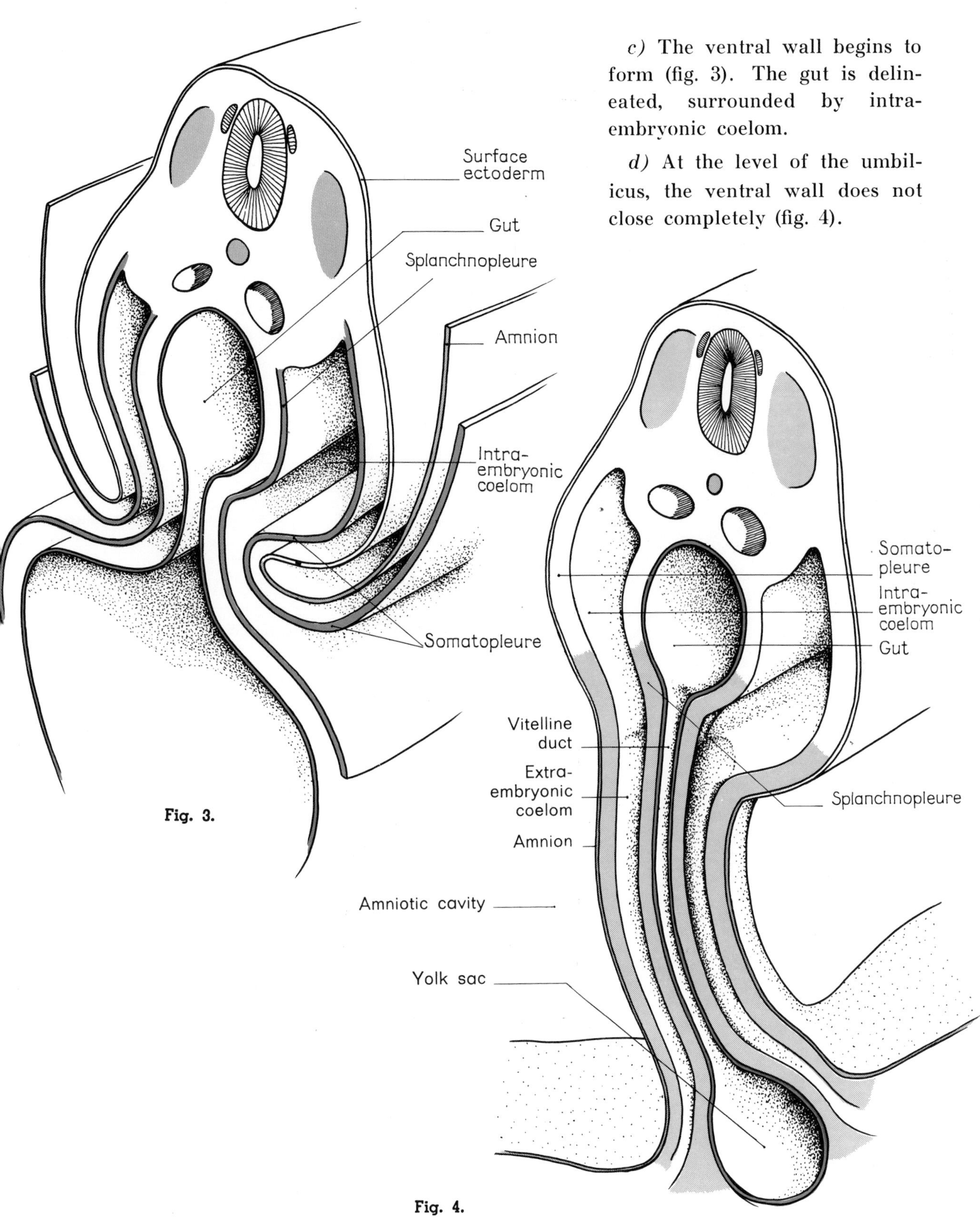

Fig. 3.

Fig. 4.

The initial processes of flexion are exactly the same above and below the level of the umbilicus as in the umbilical region. They are therefore not shown here. Eventually they result in complete closure of the body wall and formation of a ventral mesentery. Varied development of mesenteries imparts specificity to each region.

II. — REGION CAUDAL TO UMBILICUS

a) The ventral wall is closed, the gut is connected to it by a temporary ventral mesentery (fig. 1). The dorsal mesentery has not yet differentiated.

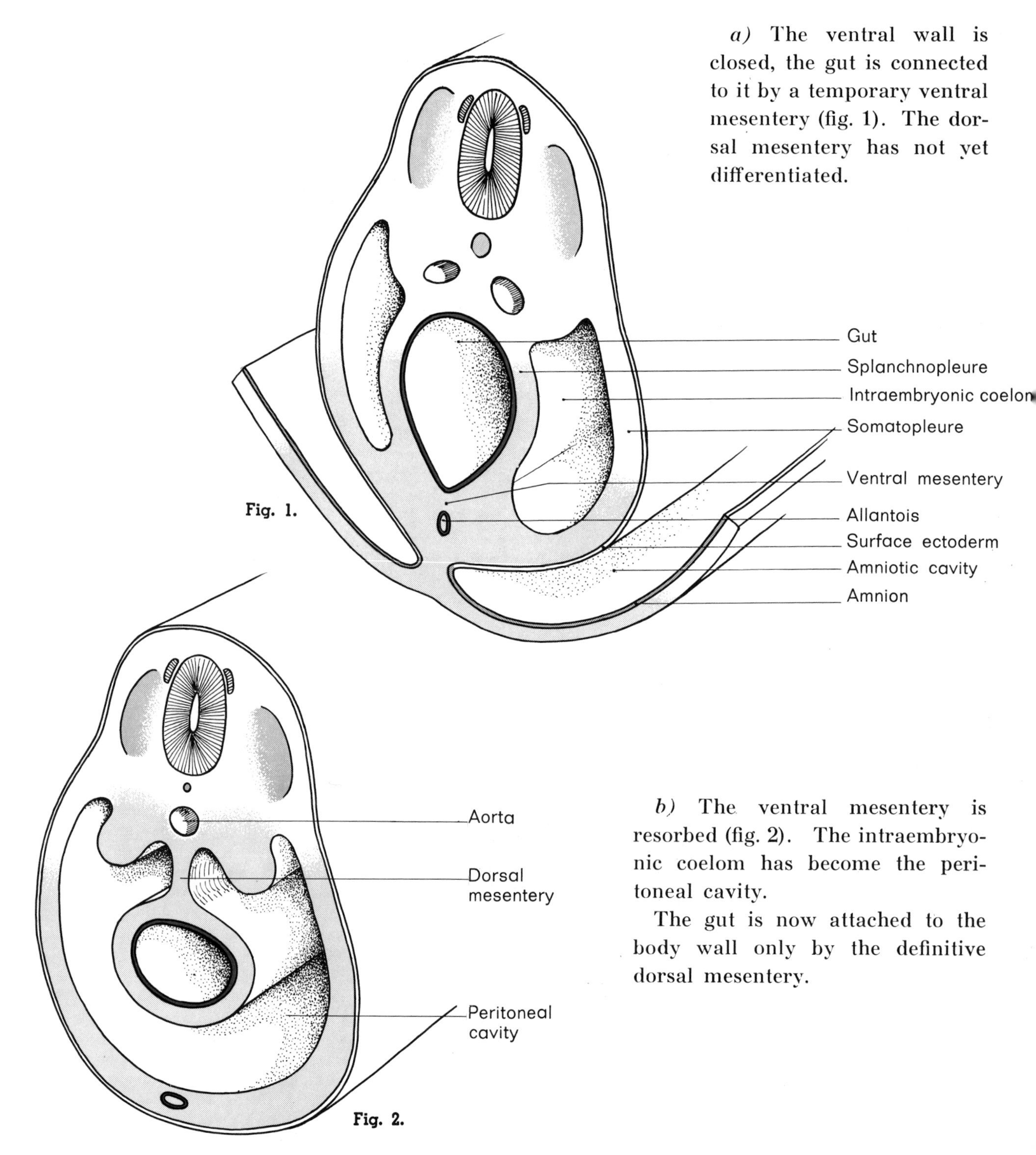

Fig. 1.

Fig. 2.

b) The ventral mesentery is resorbed (fig. 2). The intraembryonic coelom has become the peritoneal cavity.

The gut is now attached to the body wall only by the definitive dorsal mesentery.

III. — REGION CEPHALIC TO UMBILICUS

a) The section goes through the duodenal region (fig. 3). At this level, the entodermal lining gives rise dorsally to one of the pancreatic primordia, and ventrally to the liver primordium.

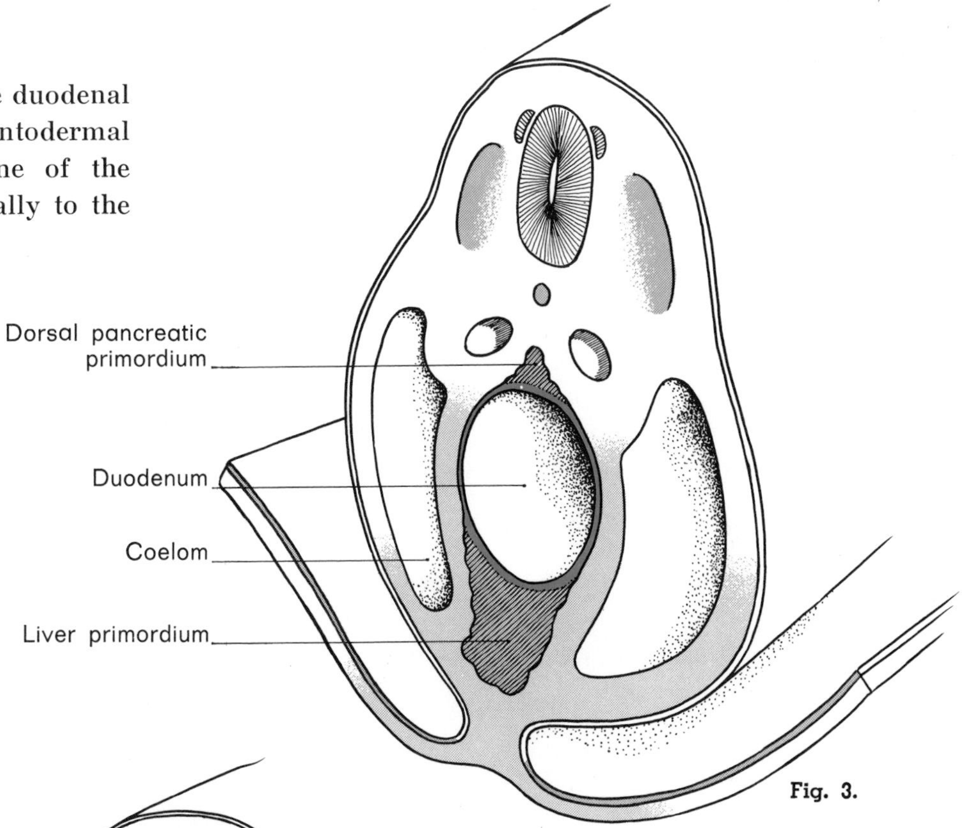

Dorsal pancreatic primordium

Duodenum

Coelom

Liver primordium

Fig. 3.

Dorsal pancreatic primordium

Peritoneal cavity

Liver

Fig. 4.

b) Extensive development of the liver primordium within the ventral mesentery leads, at this level, to partial filling of the coelomic cavity (fig. 4).

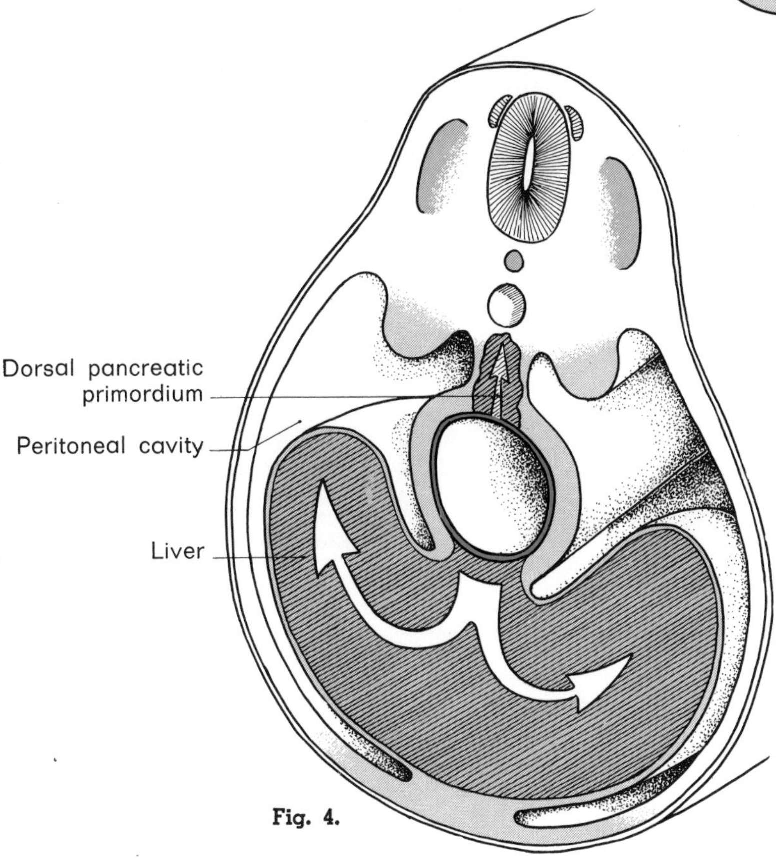

CRANIAL REGION

Modeling of the cranial end results from :

— the predominant development of the brain;
— the swinging movement of the cardiac primordium which, from its initial clearly cranial position, becomes distinctly ventral.

Brain development and cardiac movement contribute to delineation on both sides of the buccopharyngeal membrane of :

— behind the membrane, an entodermal diverticulum, the future foregut;
— in front of the membrane, a depression in the surface ectoderm open into the amniotic cavity : the stomodeum or primitive mouth.

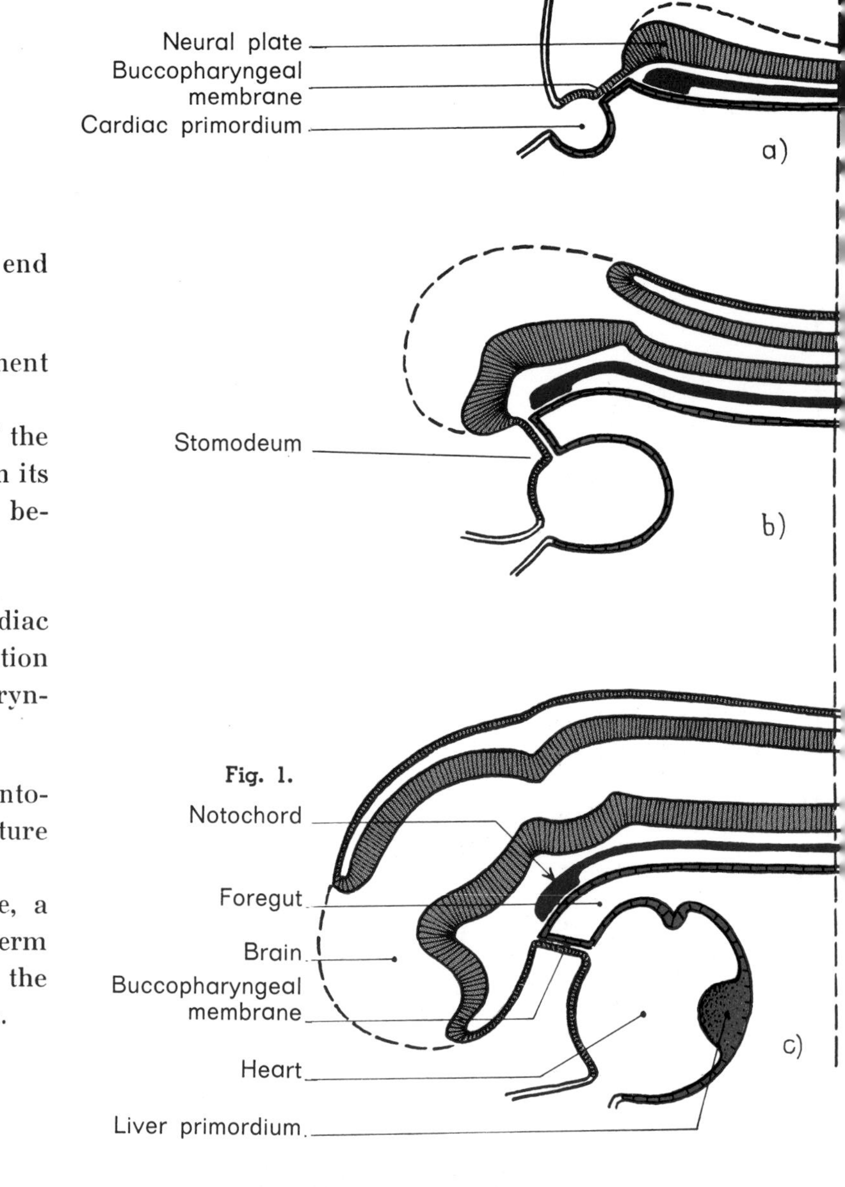

Fig. 1.

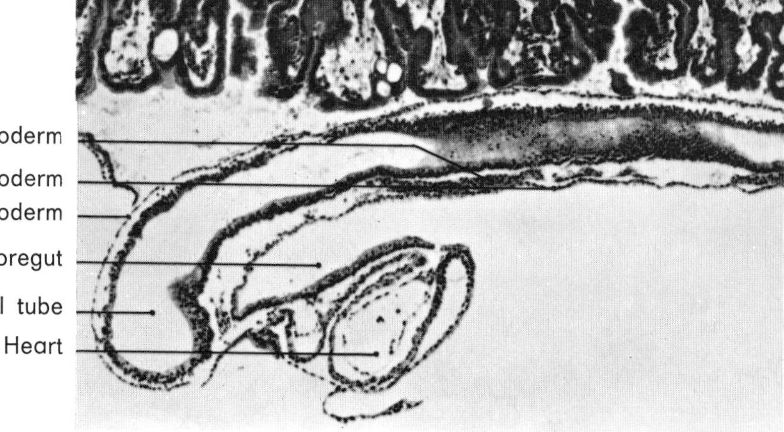

Fig. 2. — *Rabbit embryo in process of flexion.* Cranial end. Sagittal section corresponding to figure 1, *c*.

IN LONGITUDINAL SECTION

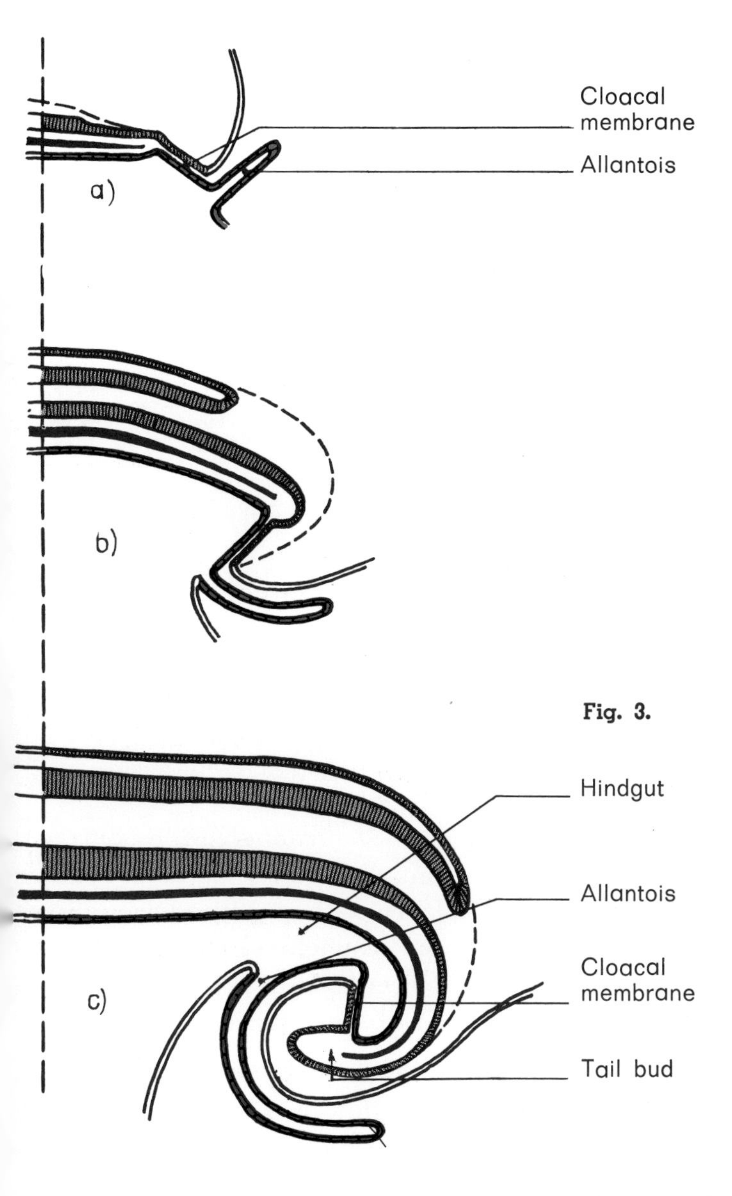

Fig. 3.

CAUDAL REGION

The caudal end of the embryo, while undergoing less development than the cranial end, goes through the same ventral curving.

The connecting stalk containing the allantois and the placental vessels passes from a primitively dorsal position to its definitively ventral position.

This movement contributes to delineation of :—

— *the cloacal region,* where the terminal part of the hindgut and the allantois meet.

— *the tail bud* containing the terminal part of the notochord, mesenchyme, and a diverticulum of the hindgut. This caudal prominence later regresses completely.

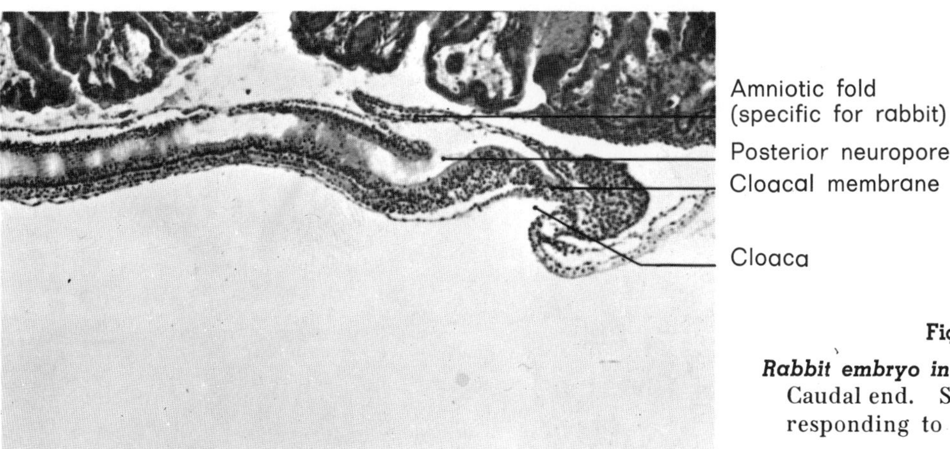

Fig. 4.

Rabbit embryo in process of flexion. Caudal end. Sagittal section corresponding to figure 3, *c*.

DEVELOPMENT

TIME OF APPEARANCE OF MEMBRANES IN THE HUMAN EMBRYO	
Trophoblast.........	5 days.
Amnion.............	7 days.
Primitive yolk sac...	9 days.
Coelom.............	12 days.
Allantois............	16 days.

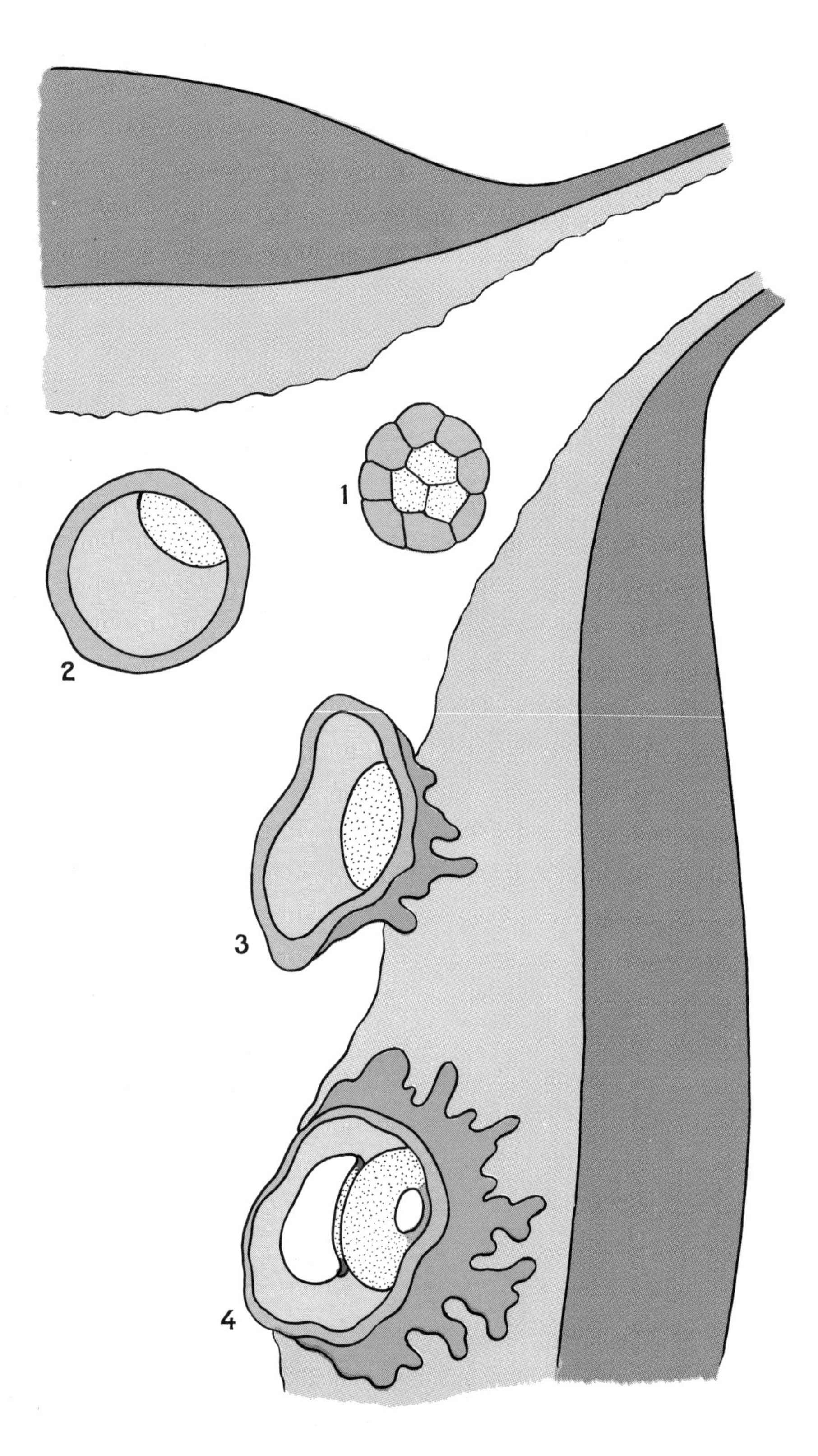

I. — FIRST WEEK

During the first week of development, the following structures appear in succession (fig. 1) :—

— *the trophoblast,* by differentiation of the superficial cellular layer of the fertilized egg. (Since the trophoblast gives rise to the placenta, it will be studied with this organ, pp. 62 to 85.)

— *the amniotic cavity,* which is hollowed out from the middle of the inner cell mass (embryoblast).

Fig. 1.

1. ***Morula stage :*** the trophoblast begins to differentiate.
2. ***Blastocyst stage*** (about 5th day); the trophoblast is differentiated.
3. ***Beginning of implantation*** (about 6th day).
4. ***Appearance of the amniotic cavity and beginning of primitive yolk sac*** bounded by entoderm and Heuser's membrane (about 7th day).

OF FETAL MEMBRANES

II. — SECOND AND THIRD WEEKS

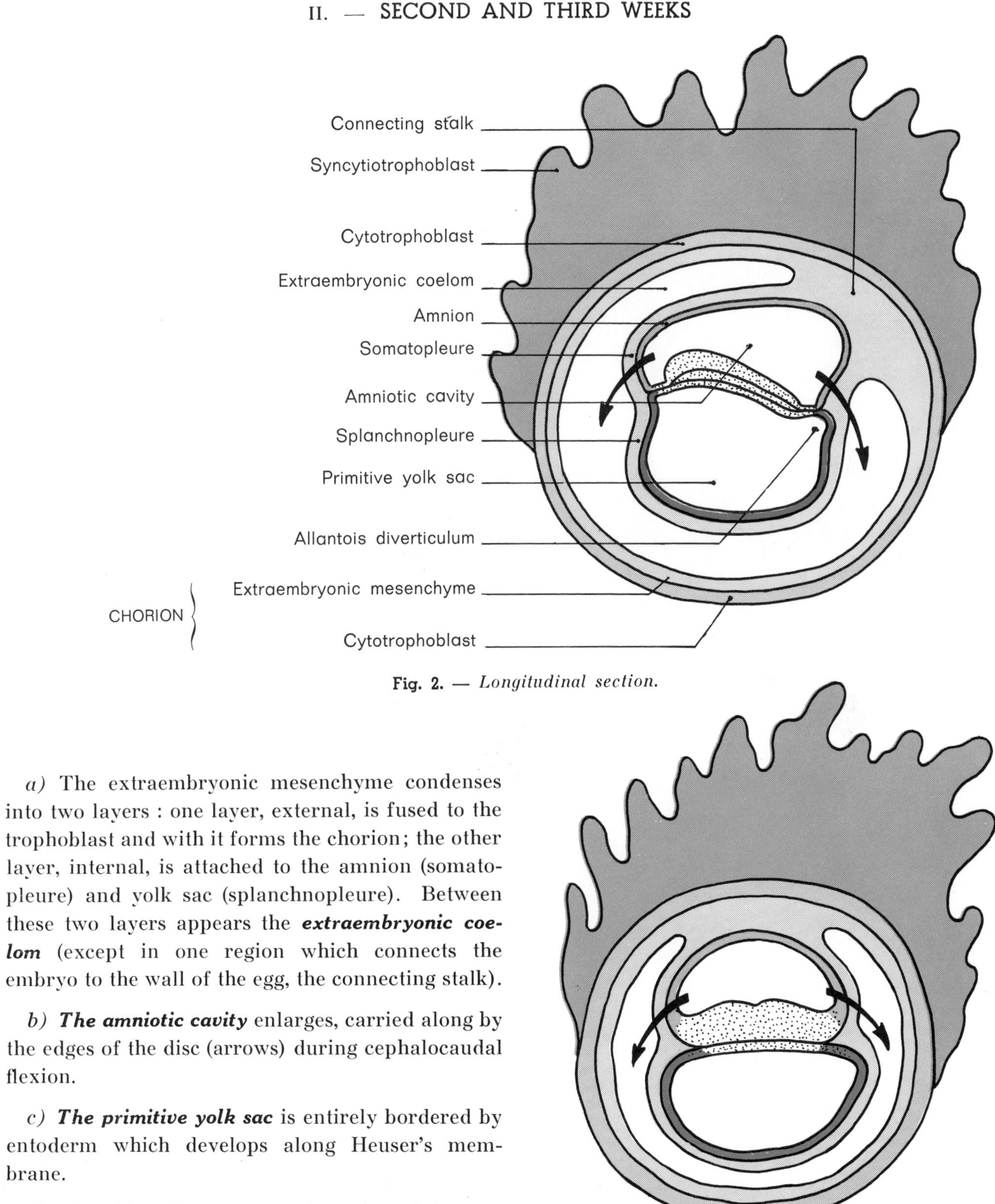

Fig. 2. — *Longitudinal section.*

a) The extraembryonic mesenchyme condenses into two layers : one layer, external, is fused to the trophoblast and with it forms the chorion; the other layer, internal, is attached to the amnion (somatopleure) and yolk sac (splanchnopleure). Between these two layers appears the ***extraembryonic coelom*** (except in one region which connects the embryo to the wall of the egg, the connecting stalk).

b) ***The amniotic cavity*** enlarges, carried along by the edges of the disc (arrows) during cephalocaudal flexion.

c) ***The primitive yolk sac*** is entirely bordered by entoderm which develops along Heuser's membrane.

d) ***The allantois*** appears at the union of the caudal part of the disc and the primitive yolk sac.

Fig. 3. — *Cross section.*

III. — FOURTH WEEK

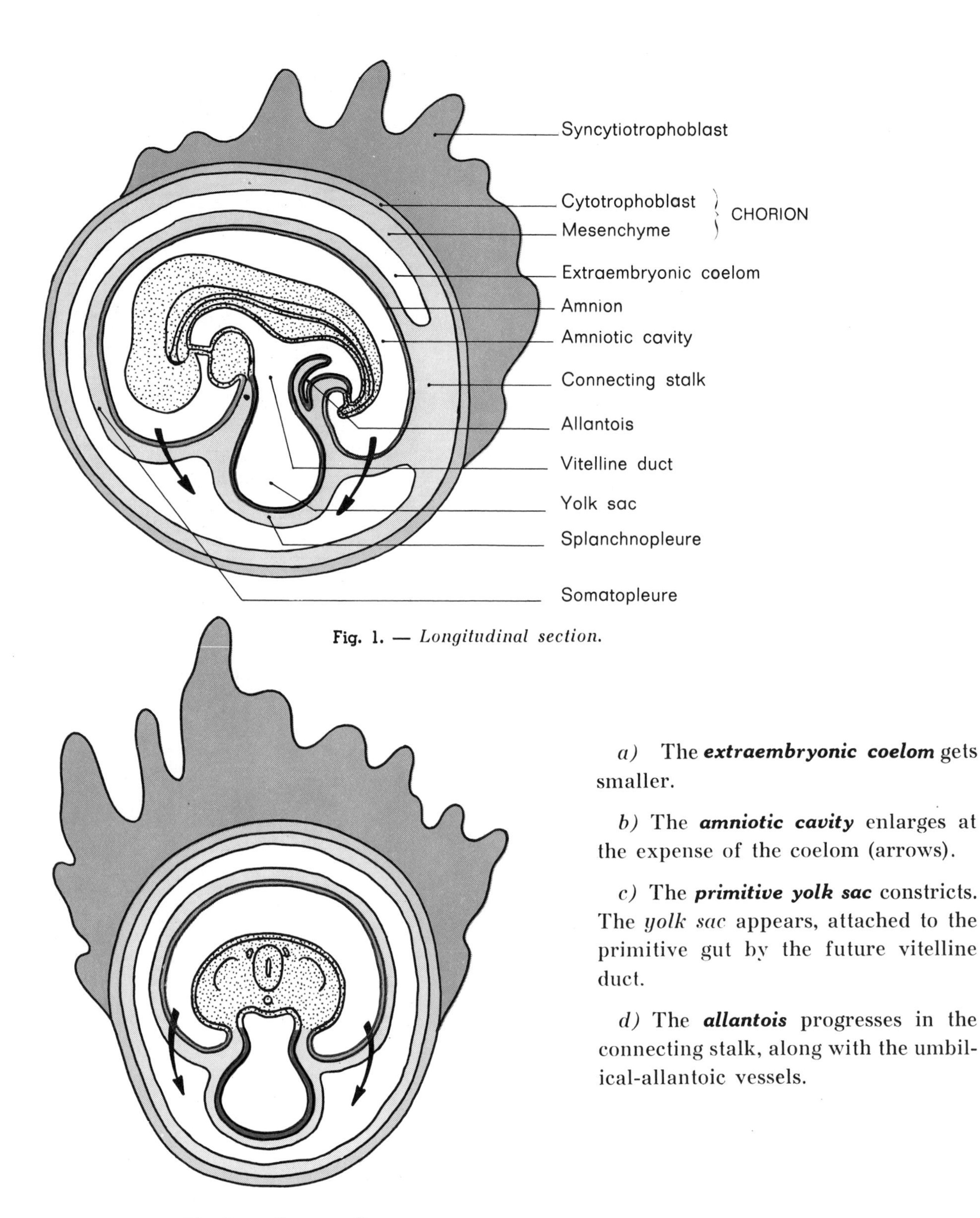

Fig. 1. — *Longitudinal section.*

Fig. 2. — *Cross section.*

a) The ***extraembryonic coelom*** gets smaller.

b) The ***amniotic cavity*** enlarges at the expense of the coelom (arrows).

c) The ***primitive yolk sac*** constricts. The *yolk sac* appears, attached to the primitive gut by the future vitelline duct.

d) The ***allantois*** progresses in the connecting stalk, along with the umbilical-allantoic vessels.

IV. — EIGHTH WEEK

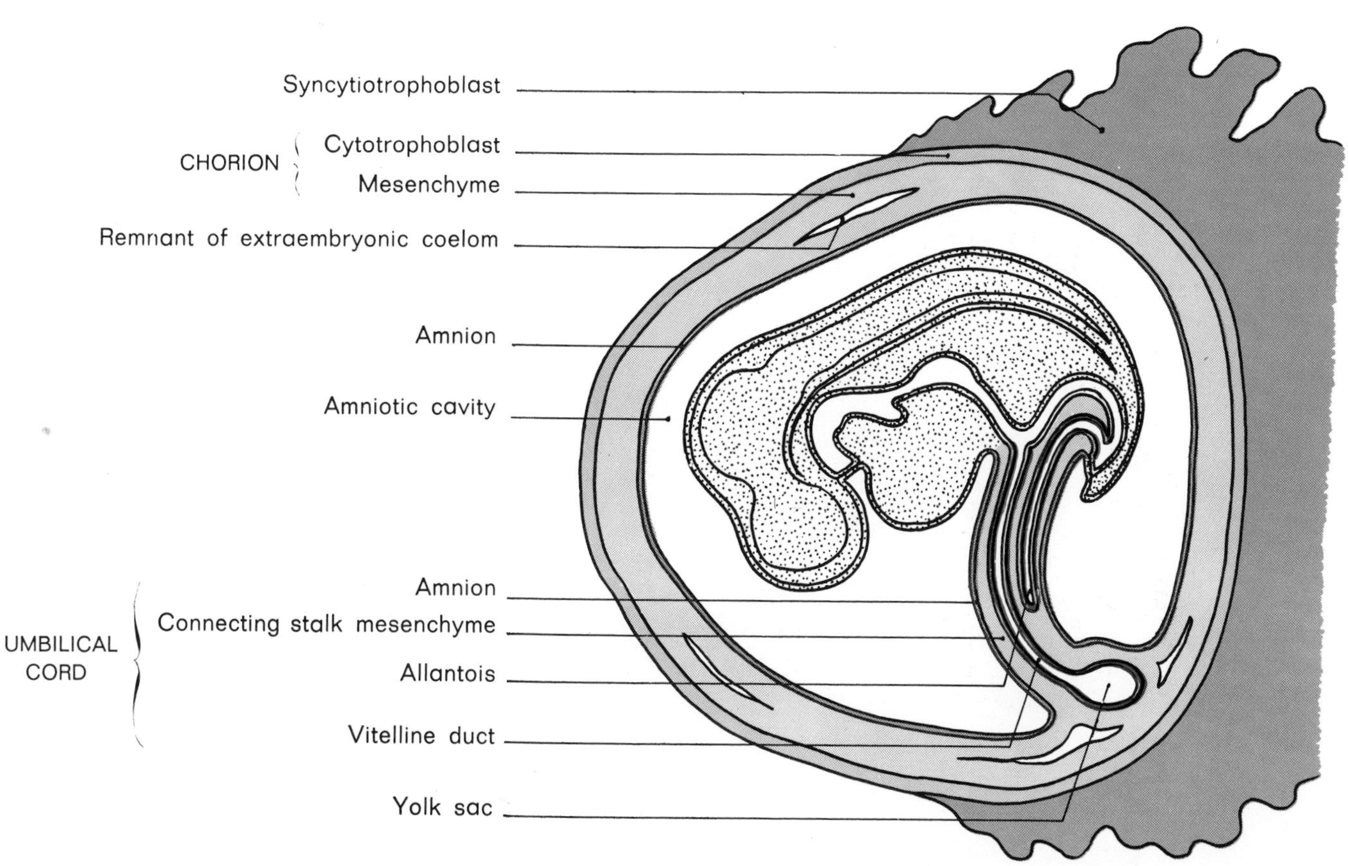

Fig. 3. — *Longitudinal section.*

a) The ***extraembryonic coelom*** disappears...

b) Effaced by development of the ***amniotic cavity.***

c) The ***yolk sac*** is right up against the placental area, at the end of the long vitelline duct which later regresses.

d) The ***allantois,*** after being extended over almost the entire length of the umbilical cord, disappears distally. The umbilical vessels continue their development.

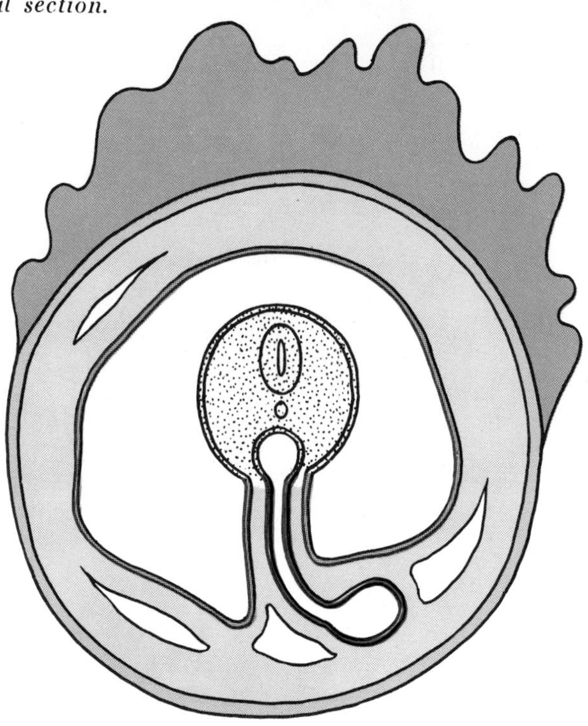

Fig. 4. — *Cross section.*

V. — FROM

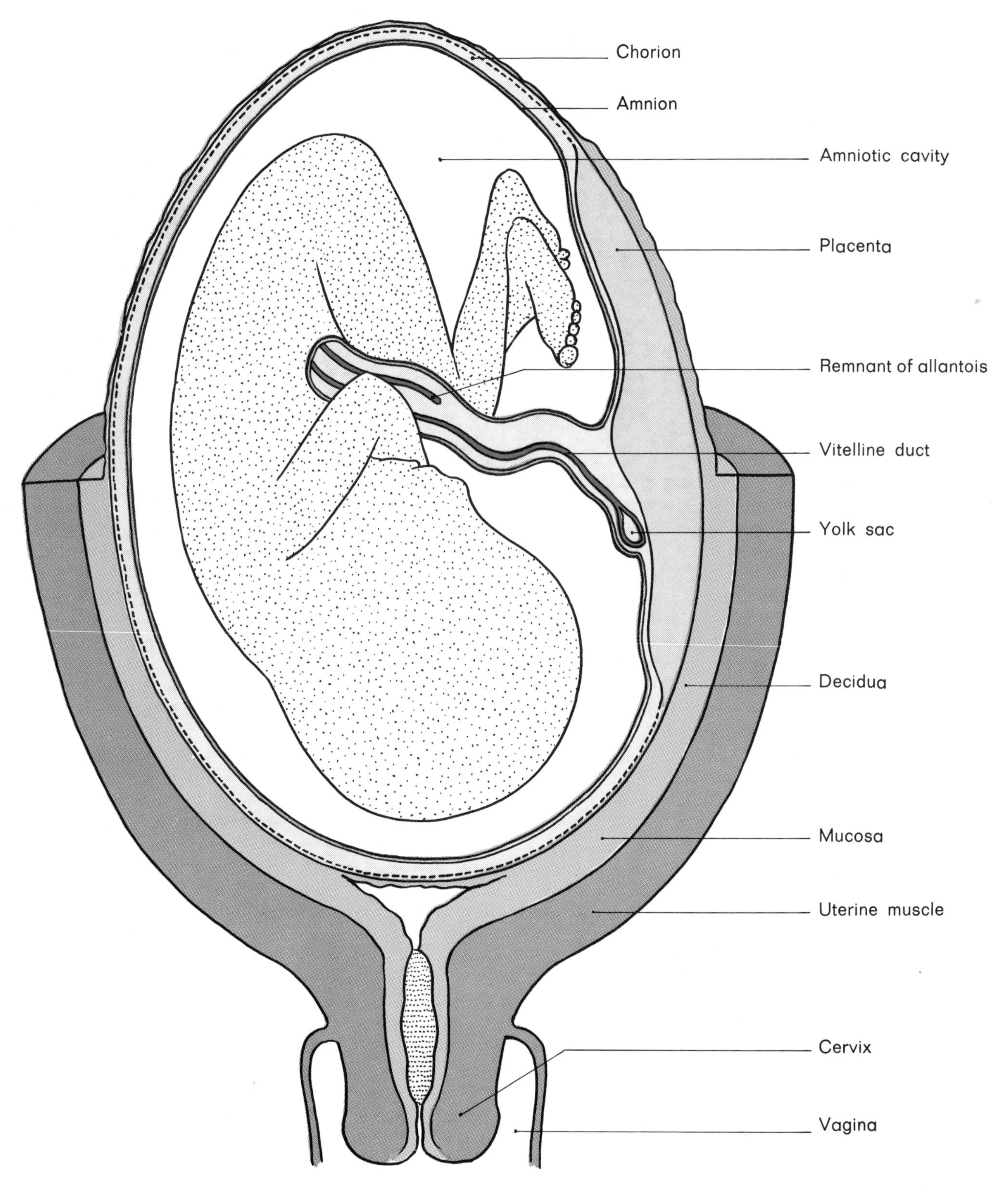

Fig. 1. — *3 1/2-month fetus and its membranes in the uterus.*

'HE THIRD MONTH

The ***yolk sac*** has almost completely :lisappeared.

The ***umbilical cord*** now contains only the umbilical vessels and also remnants of the allantois and vitelline duct.

The ***amniotic cavity*** continues to grow until term. At that time it contains about a liter of liquid : this is the so-called *bag of waters.*

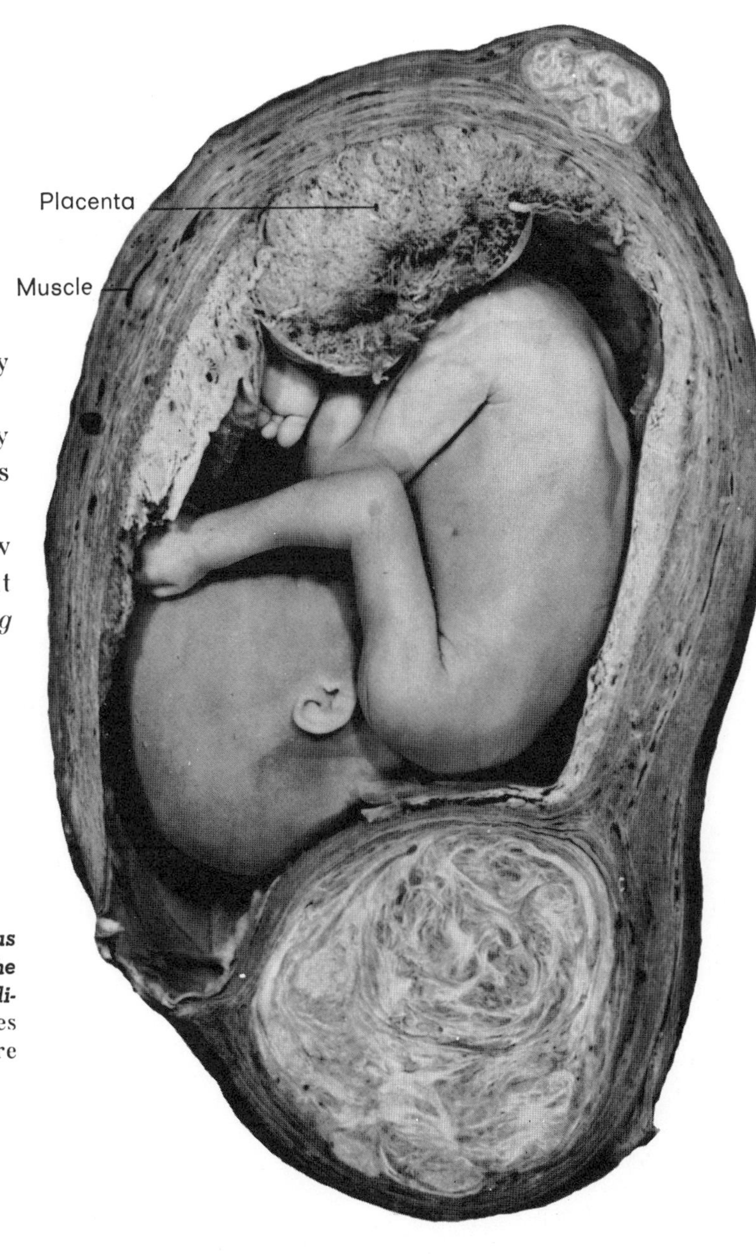

Fig. 2.

Position of a human fetus about 6 months old in the uterus sectioned longitudinally. The grayish masses seen at both poles are fibroids.

Fig. 3.

Fetal side of human placenta at term.

The membranes surround the placenta with the cord and fetal vessels in the center. The membranes are torn during labor in order to permit delivery of the infant.

VI. — THE UMBILICAL

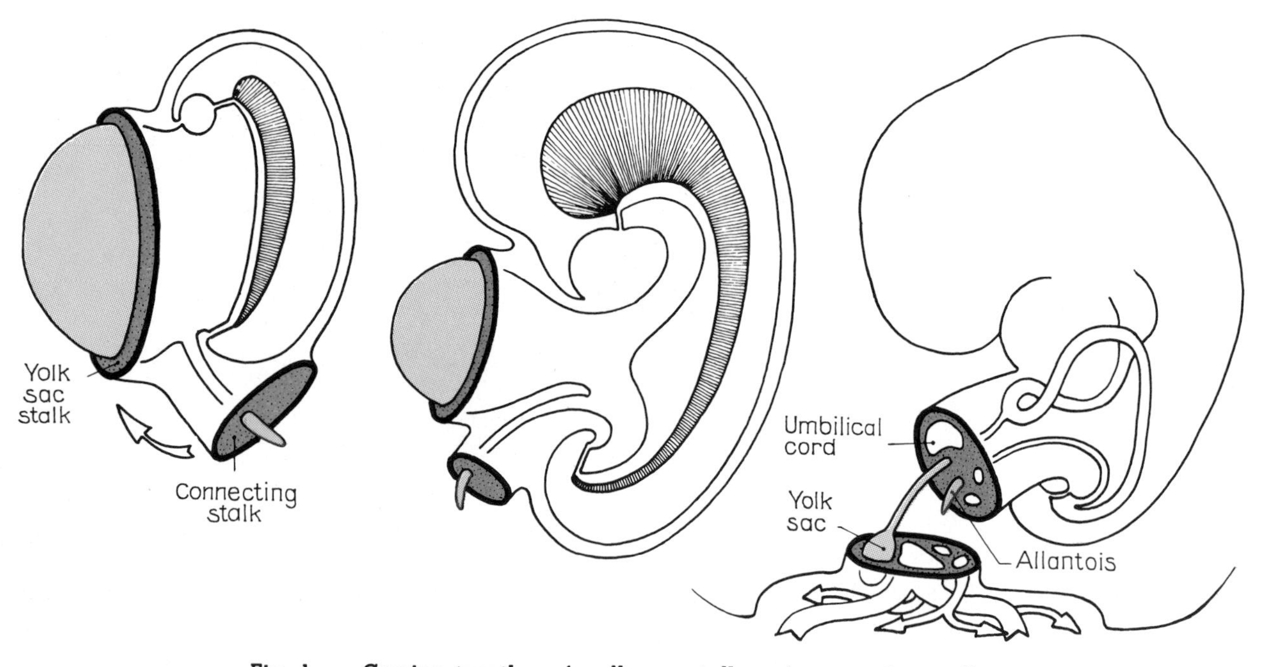

Fig. 1. — *Coming together of yolk sac stalk and connecting stalk, and formation of umbilical cord.*

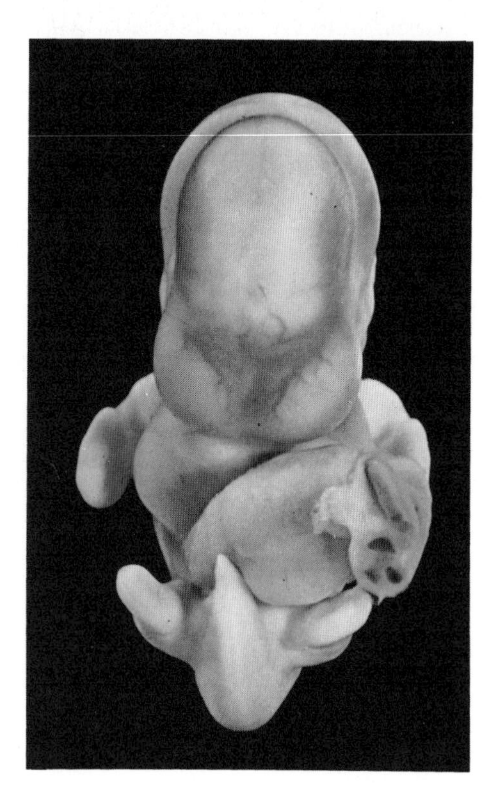

Fig. 2. — *Embryo (1 1/2 months), frontal view.*

Slightly enlarged.

At the beginning of development there are two stalks : one, clearly ventral, is *the yolk sac stalk* containing the duct and the vitelline vessels; the other, caudal, is *the connecting stalk* containing the allantois and the umbilical vessels. The latter moves ventrally as a result of cephalocaudal flexion and fuses with the yolk sac stalk to form the umbilical cord (fig. 1).

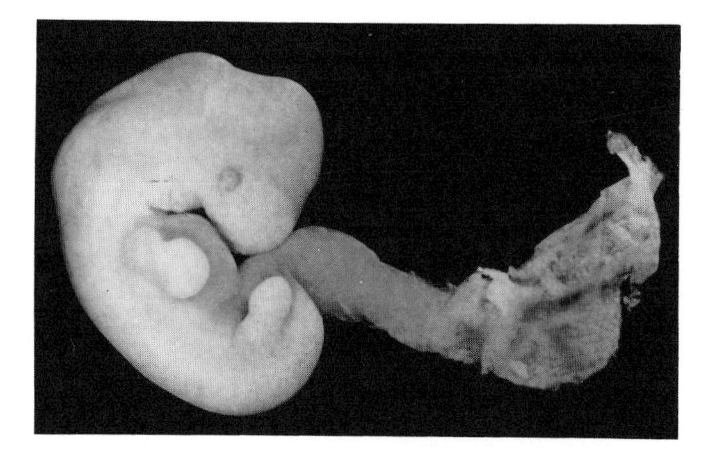

Fig. 3. — *Embryo (1 1/2 months); side view.*

Slightly enlarged.

CORD

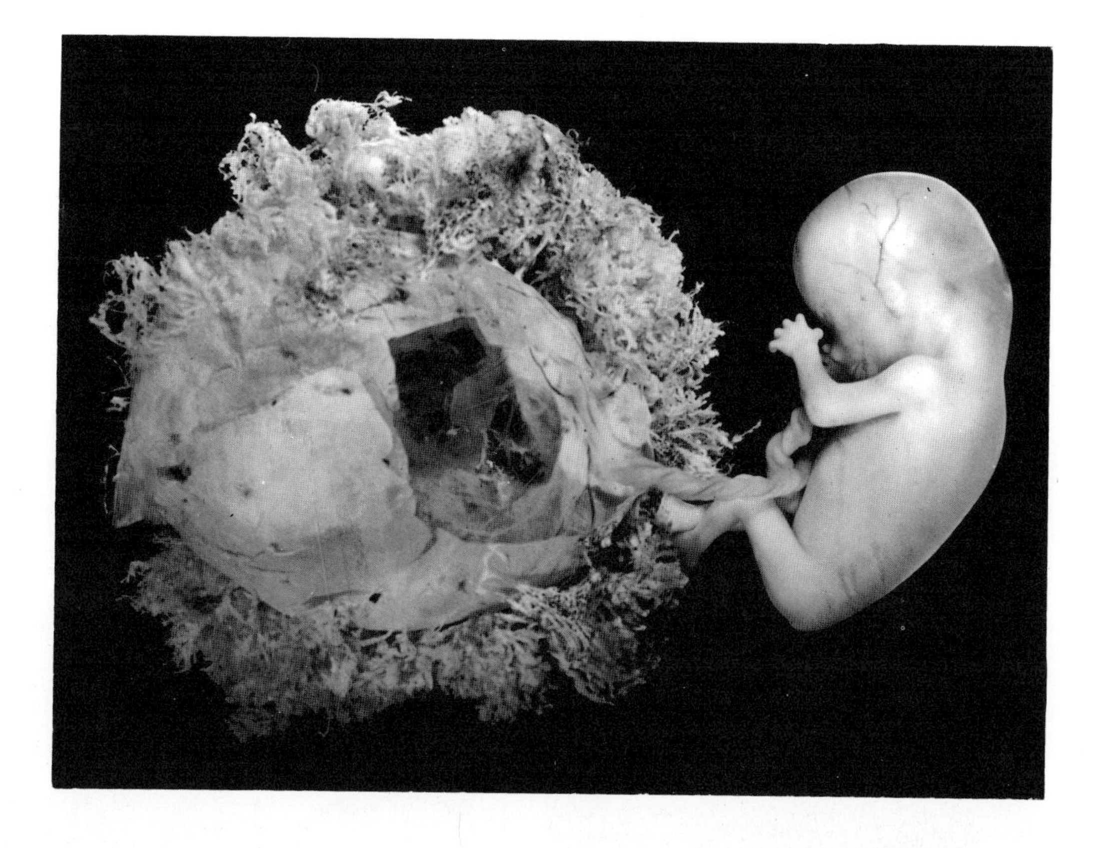

Fig. 4.
Human fetus 2 1/2 months.

The umbilical cord brings together in the same mesenchymous core, the components of the connecting stalk (allantois and umbilical vessels) and the vitelline duct.

The cord is covered by amnion which is continuous with the outer epithelial layer of the embryo at the attachment of the umbilicus.

In the young embryo, the cord is short, very thick, and is inserted in the lower part of the ventral region (fig. 1, 2, and 3).

With the development of the anterior abdominal wall, the region of umbilical implantation contracts. The umbilical cord elongates and becomes more slender (fig. 4). At term it contains only the umbilical vessels surrounded by Wharton's jelly, a smooth mesenchymous material. Its average length is then about 50 cm.

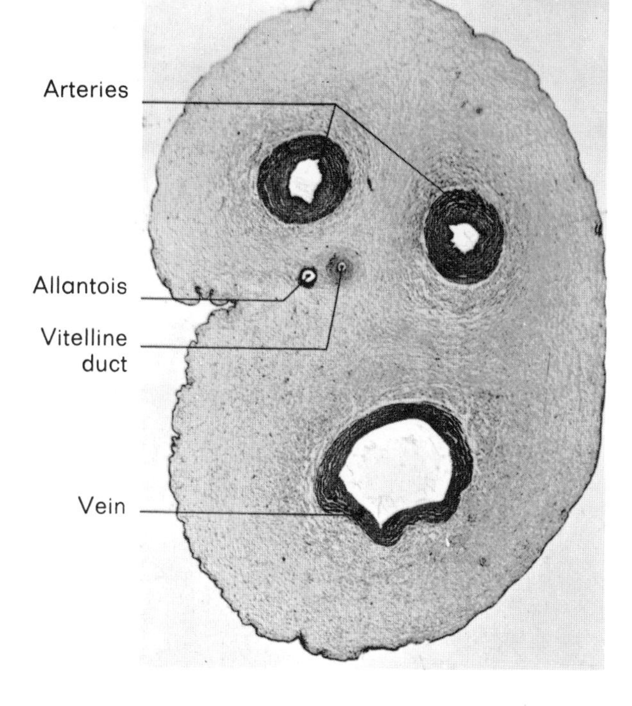

Fig. 5. — ***Umbilical cord at 3 months,*** still containing vestiges of allantois and vitelline.

THE PLACENTA :

The most important accessory fetal structure, the placenta, brings maternal and fetal blood in close relationship.

Morphologically, it is partly of fetal origin (the trophoblast), and partly of maternal origin, arising from transformation of the uterine mucosa.

The following aspects will be studied in succession :

— *exterior appearance of placenta,*
— *structure and development of the placental villus,*
— *formation of the decidua,*
— *development of vascularization of villi,*
— *overall view of fetal and maternal constituents.*

I. — EXTERIOR APPEARANCE

The trophoblast appears at 5 days. At 6-7 days it insures implantation of the egg in the mucosa because of its proteolytic activity. At this time it consists of :

— an internal cellular layer : ***the cytotrophoblast,***
— an external syncytial layer : ***the syncytiotrophoblast.***

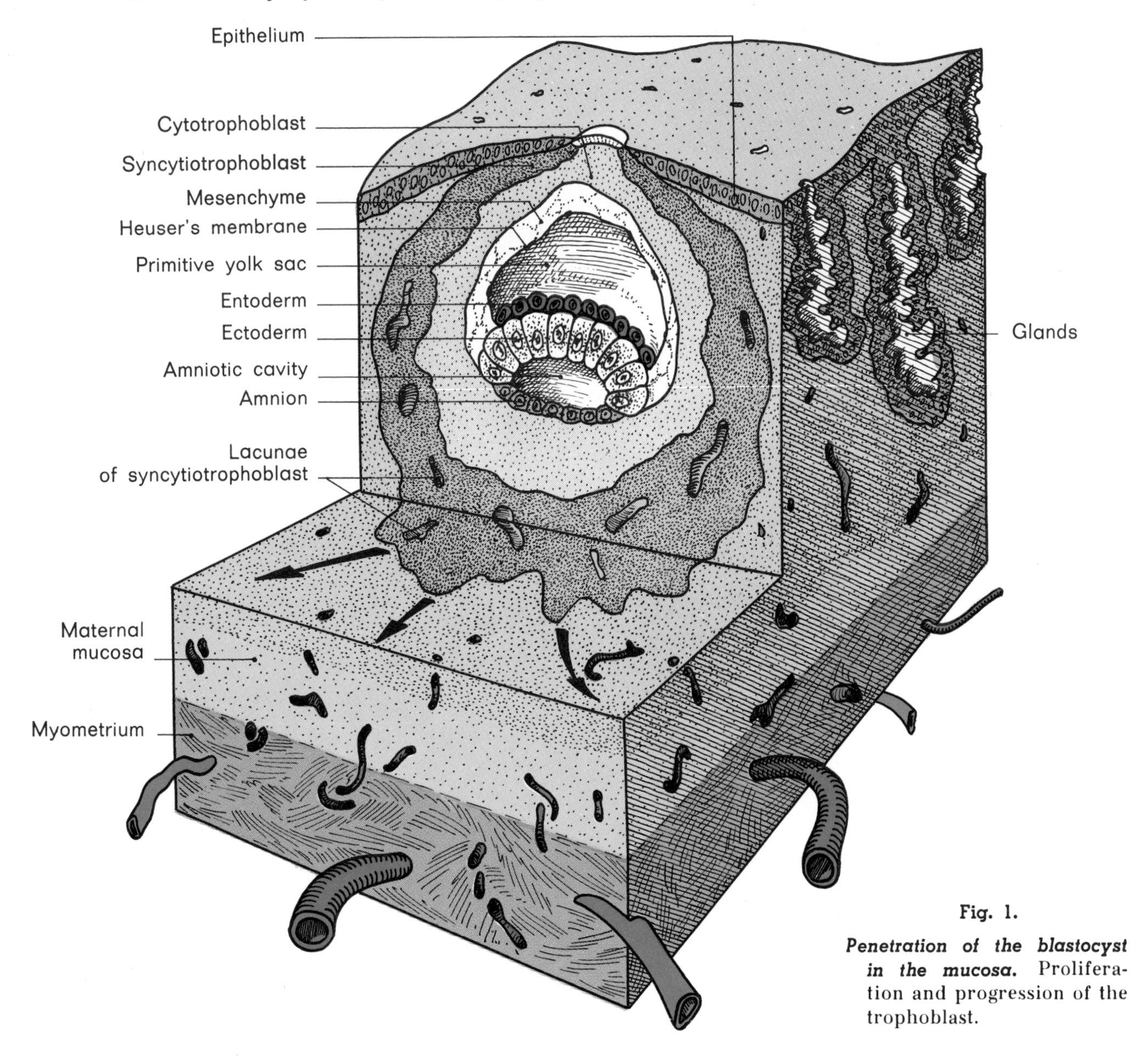

Fig. 1.

Penetration of the blastocyst in the mucosa. Proliferation and progression of the trophoblast.

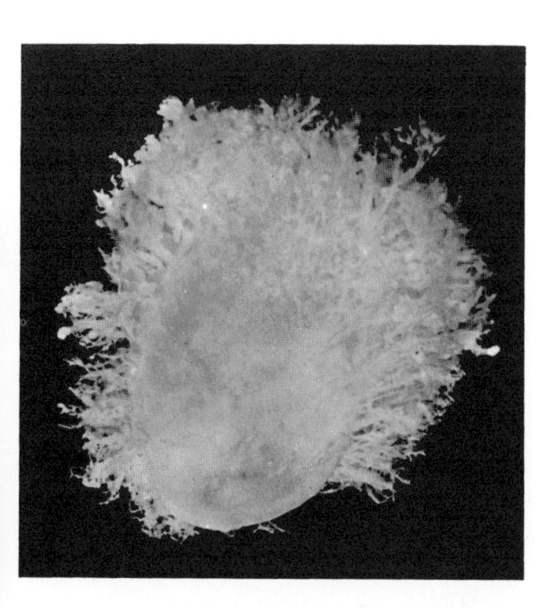

Fig. 2. — *Human ovum of about 7 weeks.* The trophoblast has proliferated into villi, visible over the entire circumference.

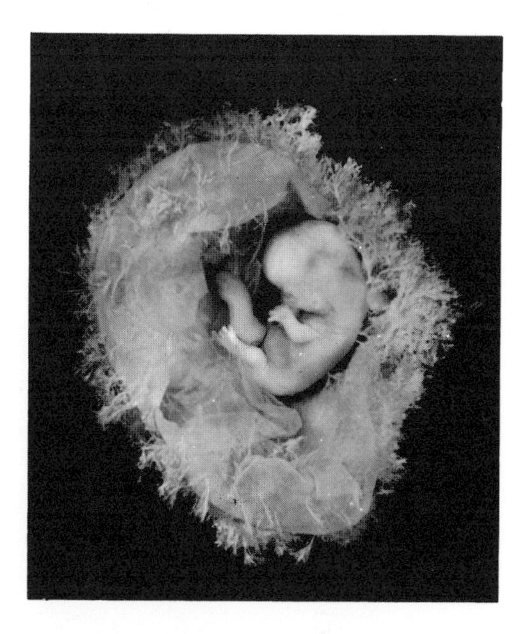

Fig. 3. — *Opened human ovum (end of 2nd month).* The villi begin to group together forming the ***chorion frondosum*** (bushy). The umbilical cord, very thick at this stage, is moving toward this region.

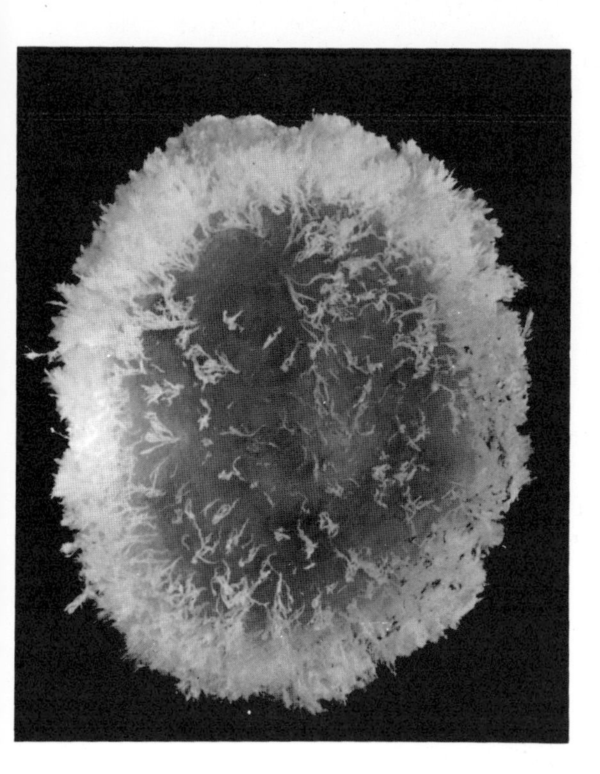

Fig. 4. — *Human ovum, 2 1/2 months.*

Rarefaction of villi
at one of the poles.

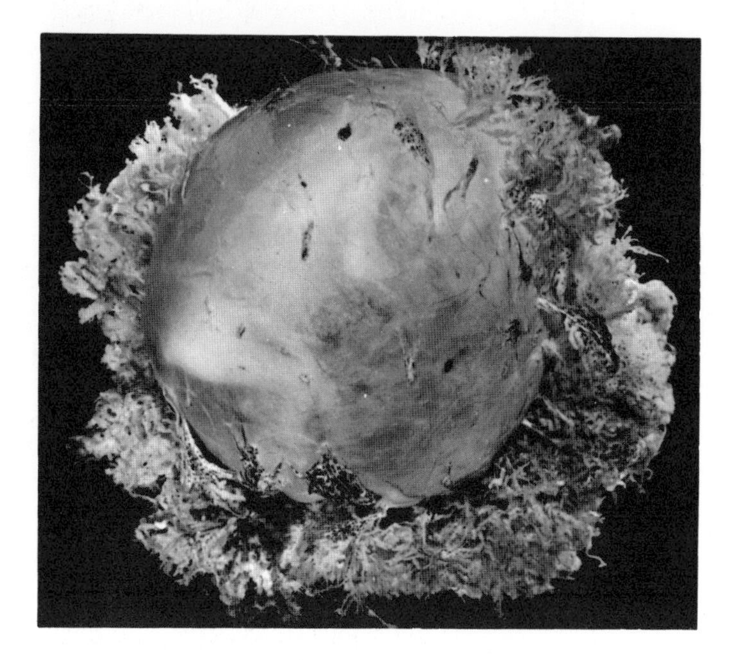

Fig. 5. — *Human 3-month egg : differentiation of the placenta.* The villi are grouped at one pole of the egg and form the placenta. The rest of the embryonic vesicle is formed by the ***chorion laeve*** (smooth) through which the outline of the fetus can be seen indistinctly.

The placenta is clearly defined by the 3rd month (fig. 1).

After this it grows, thickens and spreads out, developing along with the uterus.

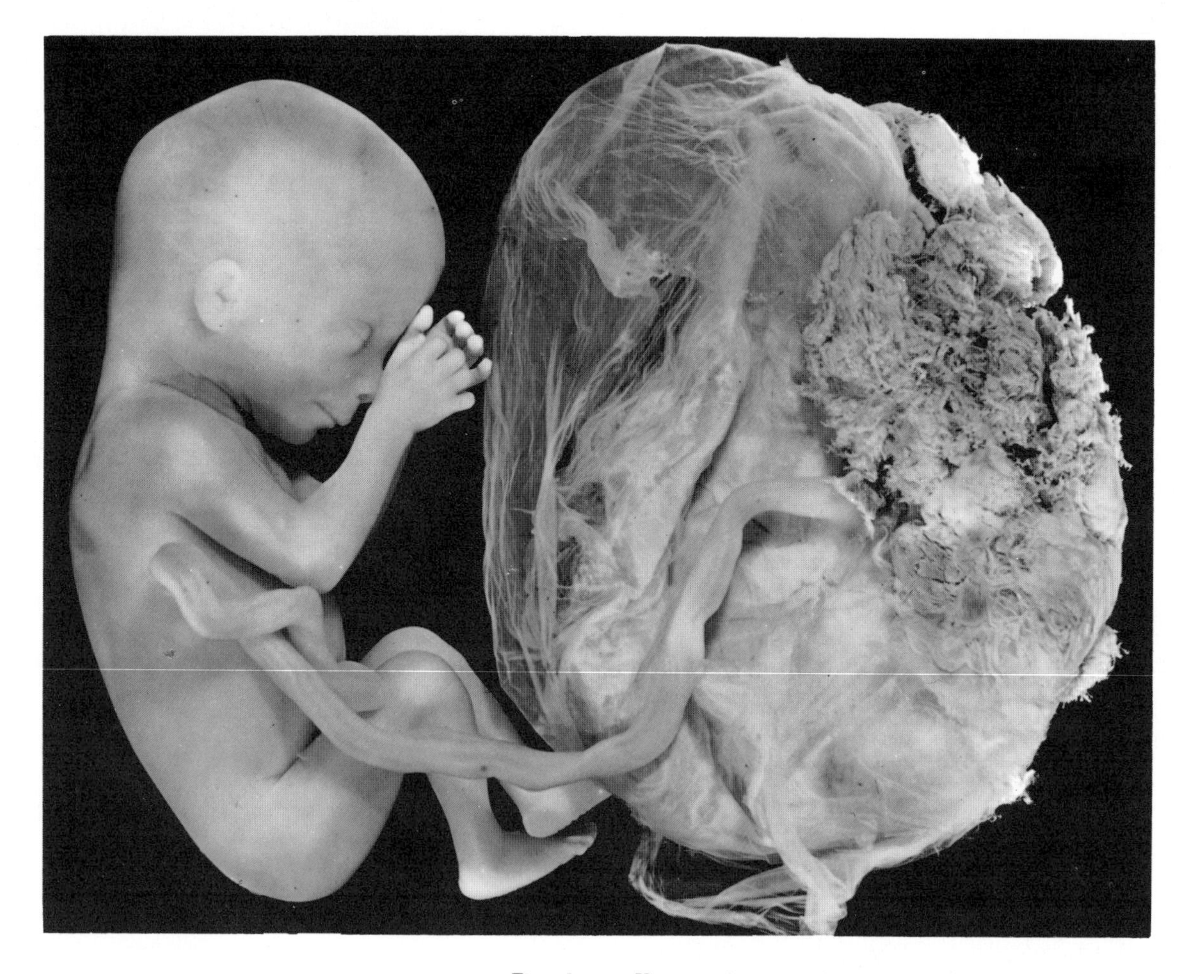

Fig. 1. — ***Human fetus and its placenta.*** The thin membrane seen beside the placenta is the amnion (about 4 months).

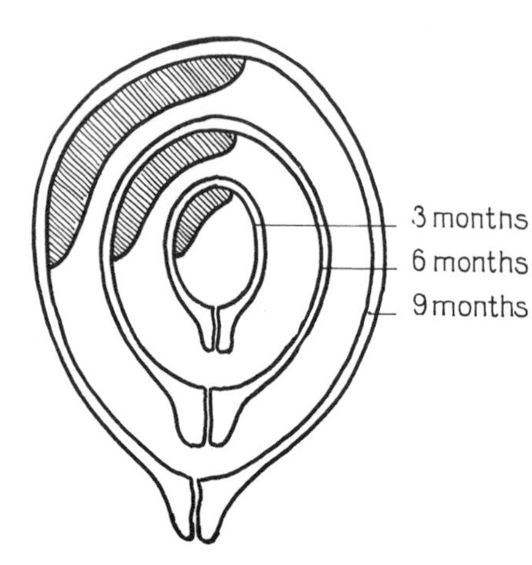

Fig. 2.

Growth of placenta in the uterine cavity.

The placenta at term is disc about 20 cm in dia-neter, 3 cm thick, and veighing about 500 g, ap-roximately 1/6 of the fetal veight. A ratio markedly lifferent from this indicates pathological condition.

Fig. 3. — *Placenta, fetal side.* The enlarged umbilical vessels are seen through the transparent amnion.

In certain cases the remnant of the yolk sac can also be seen under the amnion, at the base of the cord. This remnant is never larger than 5 mm in diameter.

In this photograph, the membranes were cut away at the edges of the placental disc (see fig. 3, p. 59, where the membranes are in place).

Fig. 4.

Placenta, maternal side.

This is formed from part of the maternal mucosa (see p. 73). Deep furrows divide the placental mass into a number of lobes or cotyledons.

II. — STRUCTURE AND DEVELOPMEN

A. — THE VILLU

The human placenta is classified as ***villous, hemochorial,*** and ***chorioallantoic*** : the placental villus is bathed directly in the maternal blood and is traversed by vessels coming from the allantoic circulation of the fetus.

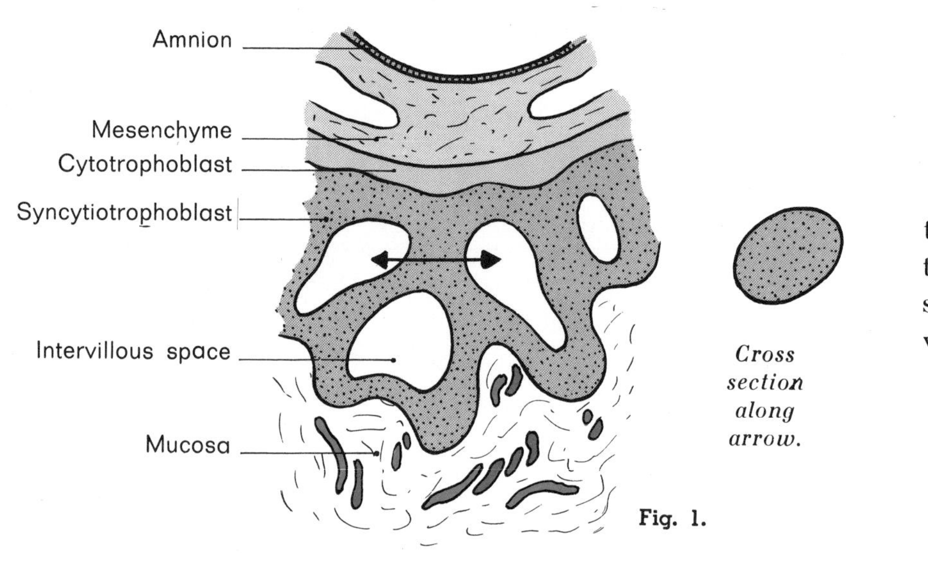

Fig. 1.

About the 13th day,

the villi begin to appear i the form of syncytial branche separated by lacunae (inter villous spaces).

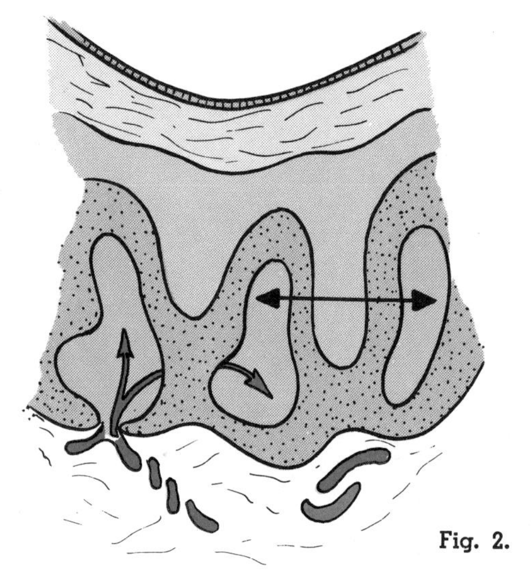

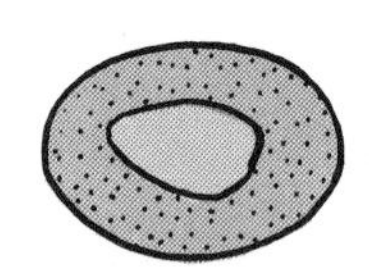

Cross section along arrow.

Fig. 2.

About the 15th day,

a cytotrophoblastic core appears inside each *cell column* with its outer covering of syncytiotrophoblast.

As the cell columns progress, they open maternal vessels whose blood spreads out in the intervillous spaces. This is the beginning of the ***maternal placental circulation.***

OF THE PLACENTAL VILLUS

BEFORE THE SECOND MONTH

About the 18th day.

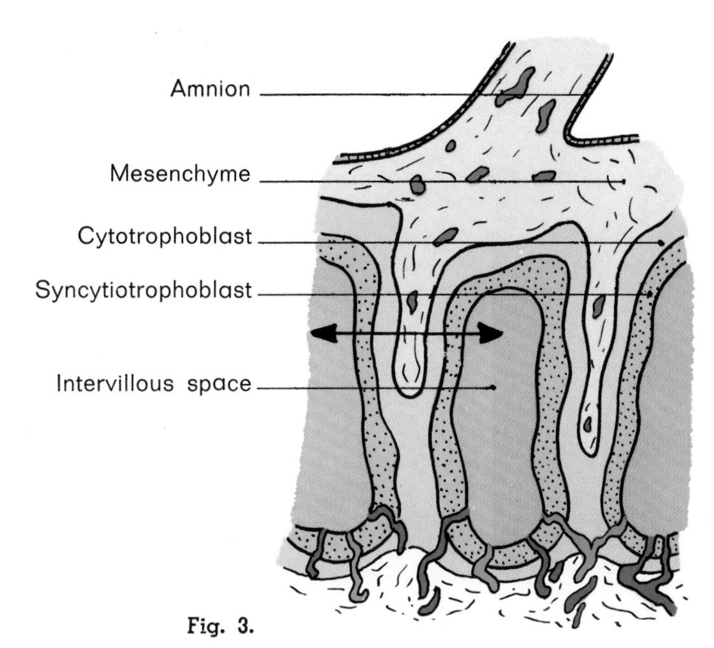

Fig. 3.

Cross section along arrow.

The villus is formed by a mesenchymal core surrounded by a double layer of cyto- and syncytiotrophoblast.

In the middle of the mesenchyme, vascular islets appear, the beginning of the future fetal circulation.

In contrast, the lacunae, now intervillous spaces, are already sites of intense maternal circulation.

About the 21st day.

The intravillus vascular network connects with the umbilical-allantoic vessels. ***Fetal placental circulation is established.***

Because of this type of circulation, using the allantoic vessels, the human placenta is termed *chorioallantoic*.

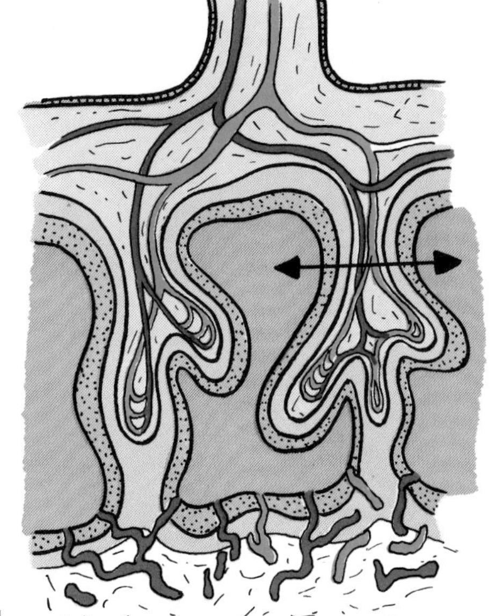

Fig. 4.

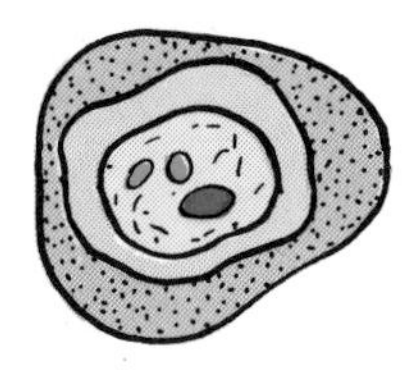

Cross section along arrow.

B. — THE VILLUS FROM THE SECOND TO THE FOURTH MONTH

The villus develops a tree-like form (fig. 1). It is bordered by a double trophoblastic layer :—

— superficial : *syncytiotrophoblast*.
– deeper : *cytotrophoblast* (cells of Langhans).

Certain branches of the villus tree make contact with the maternal tissue : *anchoring villi*. The others remain free in the intervillous spaces : *floating villi*.

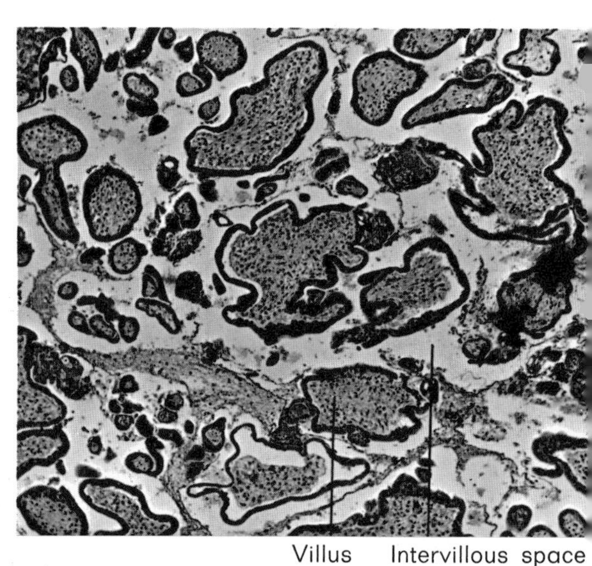

Fig. 2. — *Placenta at 2 months.*

In figure 2 *a*, note the extraordinary villu density of a placenta of more than 4 months contrasted with the appearance of a placent of 2 months (fig. 2). These two photograph were taken at the same magnification (× 62)

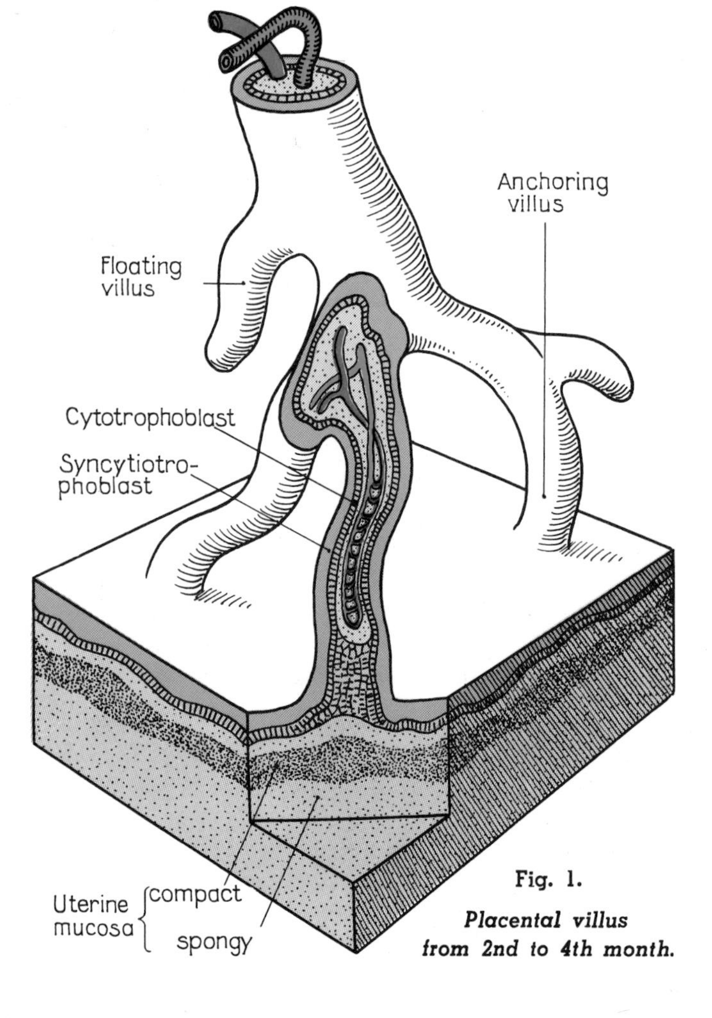

Fig. 1.
Placental villus from 2nd to 4th month.

Vessel
Syncytio-trophoblast
Connective tissue core
Cytotro-phoblast

Fig. 3. — *Villus at 2 months.*

C. — THE VILLUS AFTER THE FOURTH MONTH

Through numerous branchings the villus has become a bushy tree. Its branches form a tangled mass in whose meshes the maternal blood circulates.

The cytotrophoblast has almost completely disappeared (fig. 4).

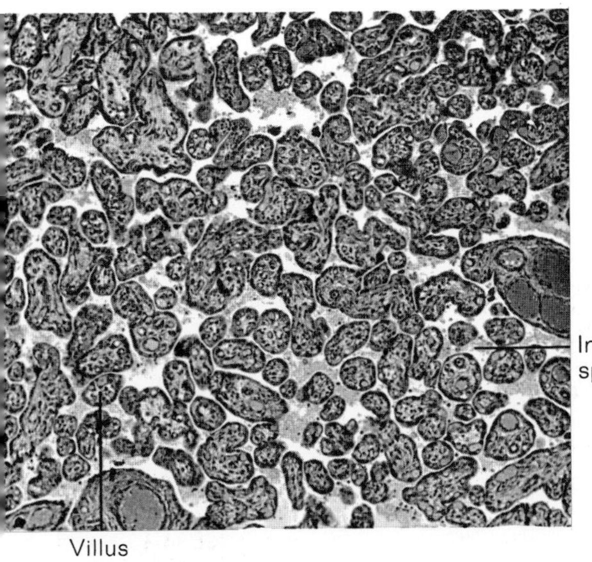

Fig. 2 *a*. — *Placenta of more than 4 months.*

In figure 3 *a* (villus after the 4th month) note the rich vascularity and thin coat (through disappearance of cytotrophoblast), contrasting with the appearance of a villus at 2 months (fig. 3). (Identical magnification : × 400.)

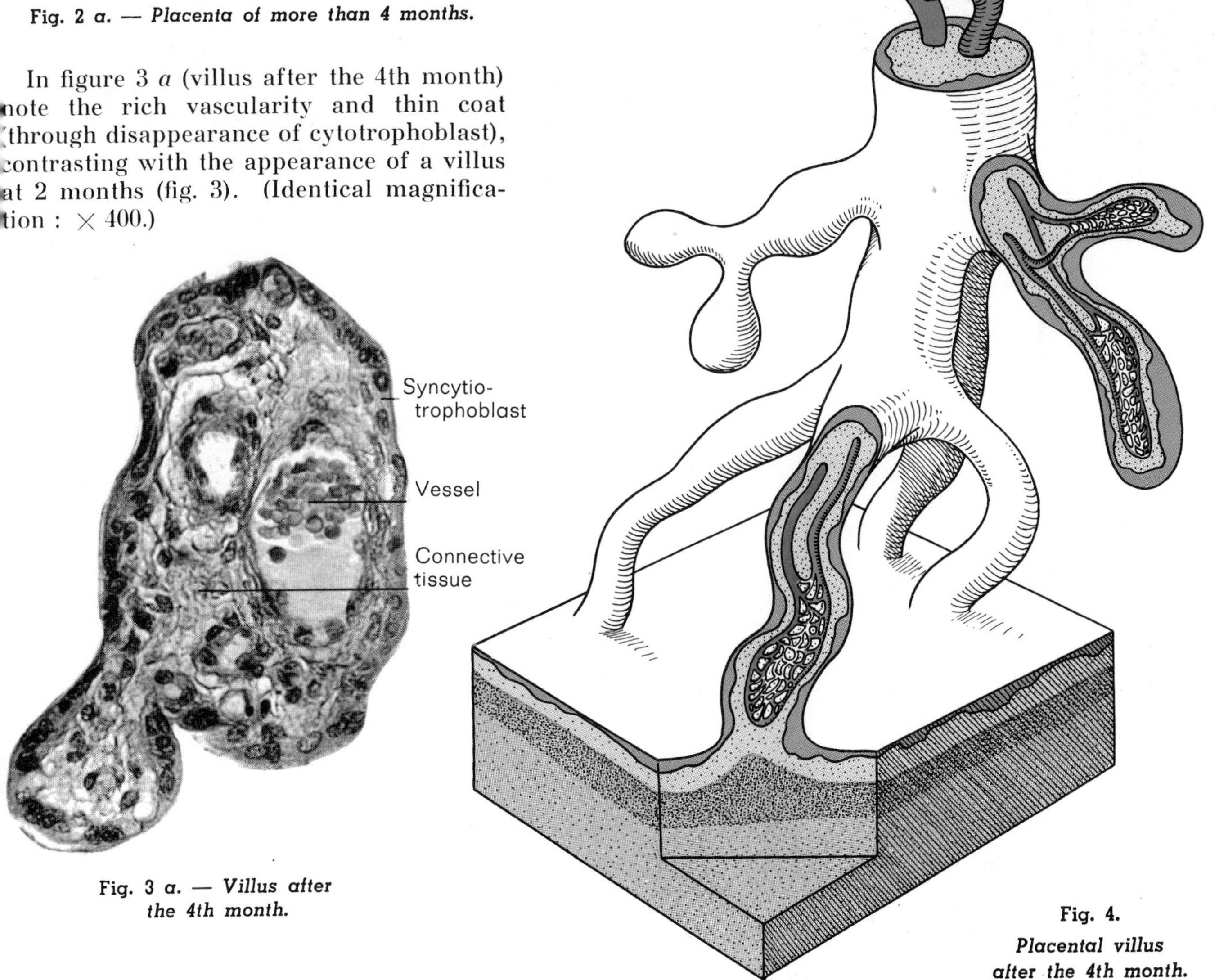

Fig. 3 *a*. — *Villus after the 4th month.*

Fig. 4.
Placental villus after the 4th month.

III. — FORMATION

The decidua result from the changes in the uterine mucosa accompanying pregnancy. There are three : basalis, parietalis, and capsularis (fig. 1).

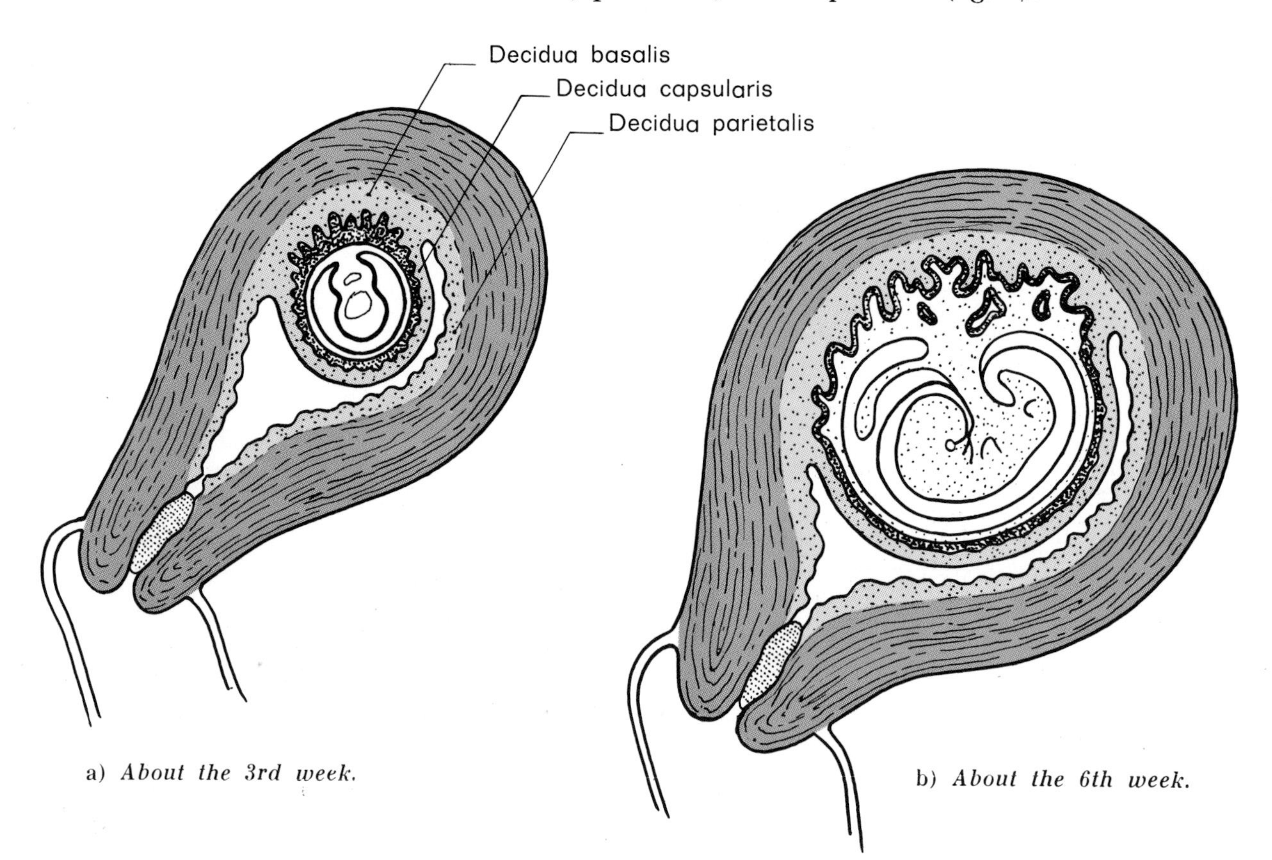

Fig. 1, *a* and *b*. — *Arrangement of the decidua during gestation.*

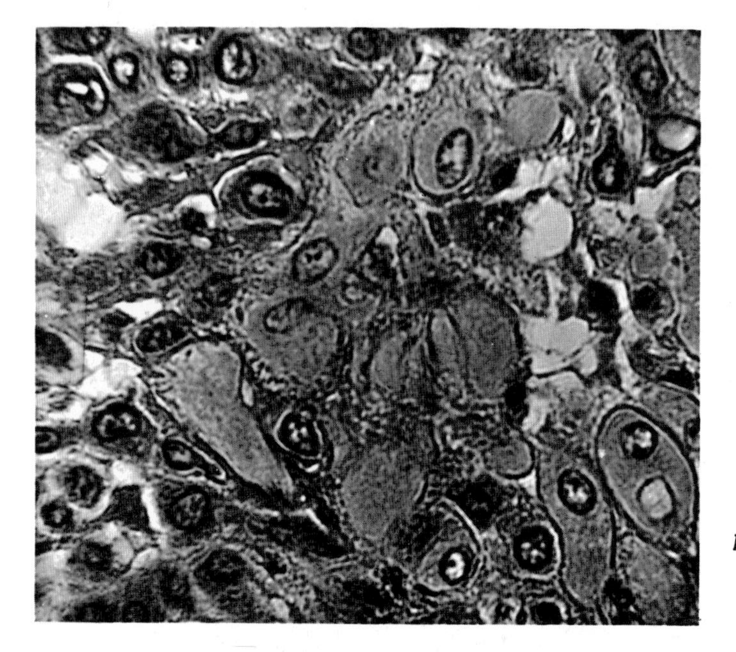

At the time of implantation, the connective tissue cells of the maternal mucosa undergo an *epithelioid* transformation.

This is the *decidual reaction* (fig. 2) which forms the compact zone of the mucosa; here no trace of the surface portion of the uterine glands is seen. In the layer just below this compact zone, the cul-de-sacs of the glands persist; this is the spongy zone through which the plane of cleavage will pass at the time of parturition (see figure 4 opposite and p. 73).

Fig. 2.
Human decidual cells* (× 500) *in the decidua basalis.

OF THE DECIDUA

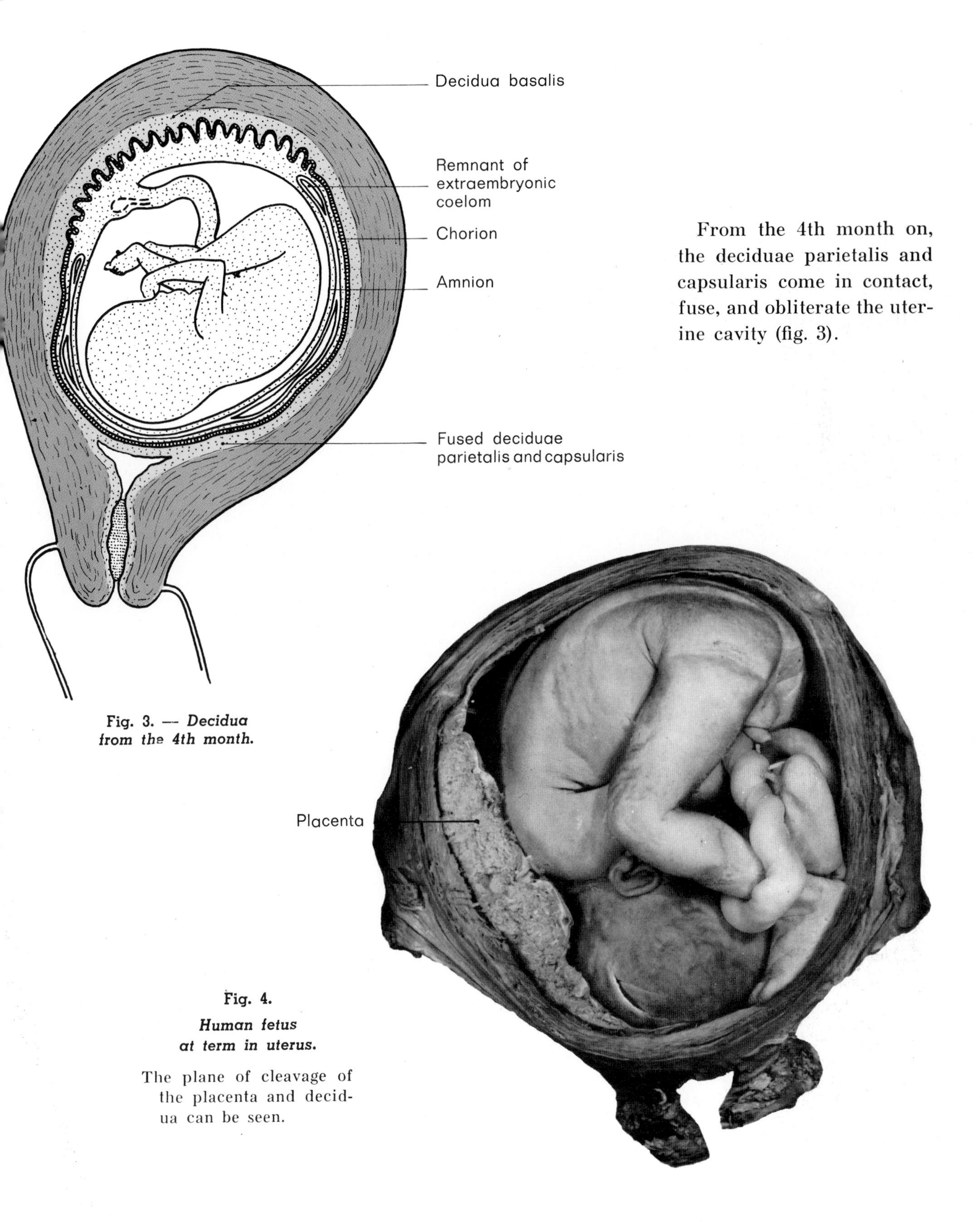

From the 4th month on, the deciduae parietalis and capsularis come in contact, fuse, and obliterate the uterine cavity (fig. 3).

Fig. 3. — *Decidua from the 4th month.*

Fig. 4.

Human fetus at term in uterus.

The plane of cleavage of the placenta and decidua can be seen.

IV. — DEVELOPMENT OF VILLUS VASCULARIZATION

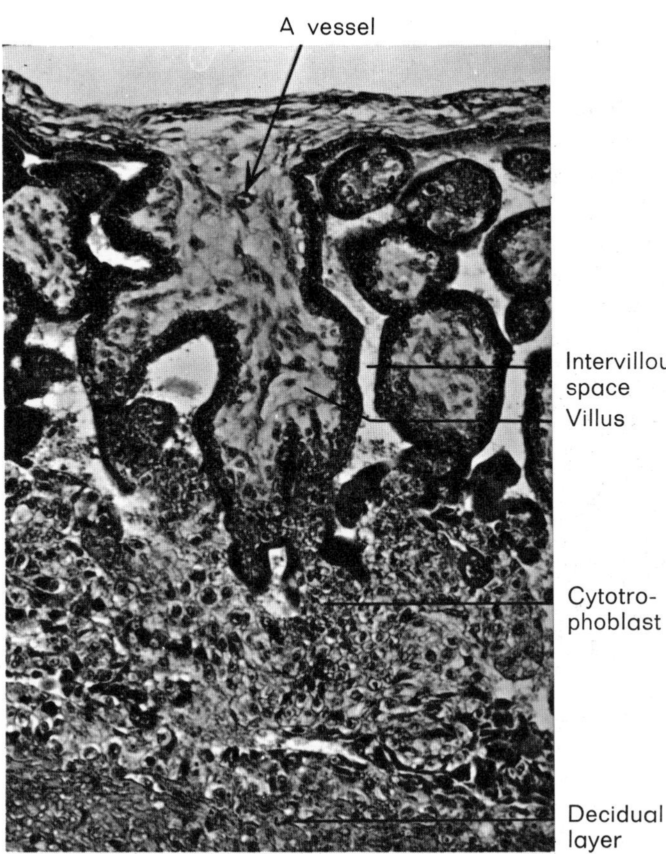

Fig. 1. — ***Placenta of one month.***
A single vessel can be seen in the villus.

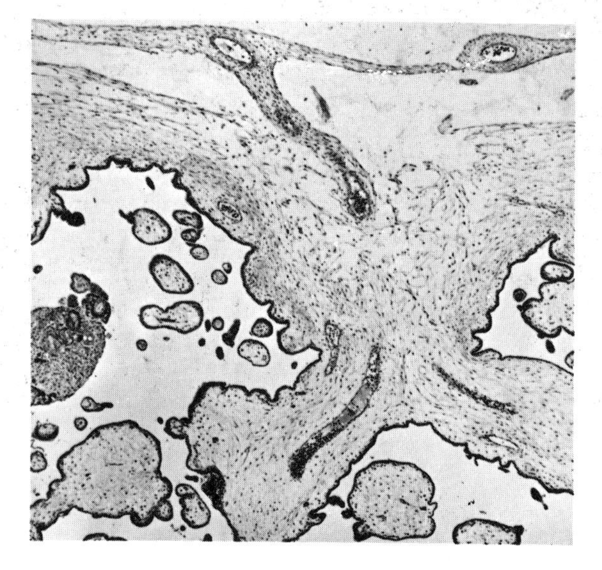

Fig. 2. — ***Placenta of 2 months.***
The vascular core has developed.

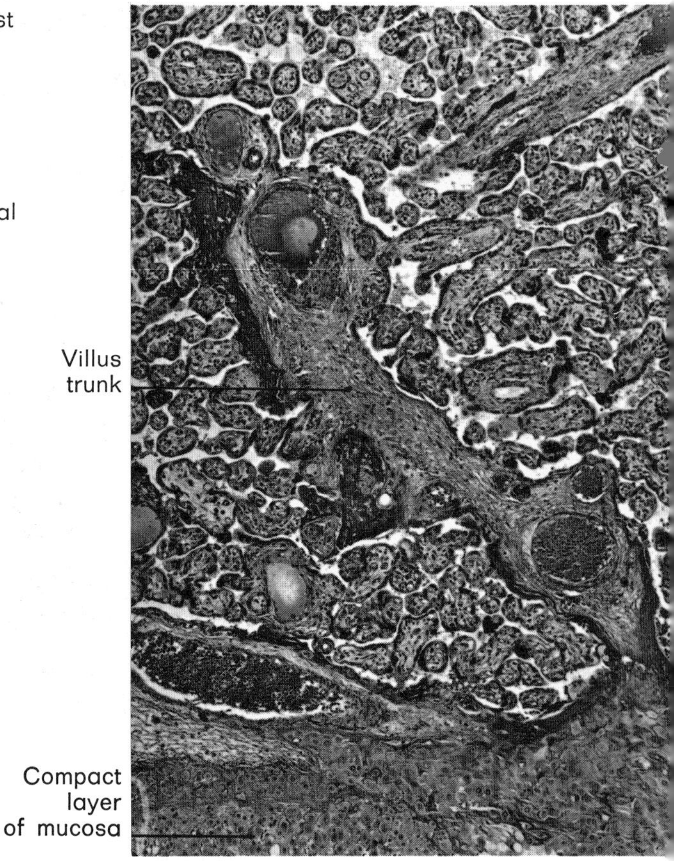

In the photomicrographs of figures 1, 2, and 3, the appearance and development of vessels in the villus trunk can be followed.

Villus trunk

Compact layer of mucosa

Fig. 3. — ***Placenta at term*** showing an anchoring villus; large vessels follow the villus core up to the basal plate. These vascular trunks give rise to the capillary networks which are involved in all the branchings.

V. — COMPOSITE VIEW OF FETAL AND MATERNAL CONSTITUENTS

There are 20 to 30 large villus trunks which correspond to the cotyledons already described (lobes seen on the maternal side of the placenta at term). Each trunk with its branchings floats in a space partitioned laterally by the decidual septa; these structures, which appear in the fourth month, start from the maternal (decidual) plate, but do not reach the chorionic plate.

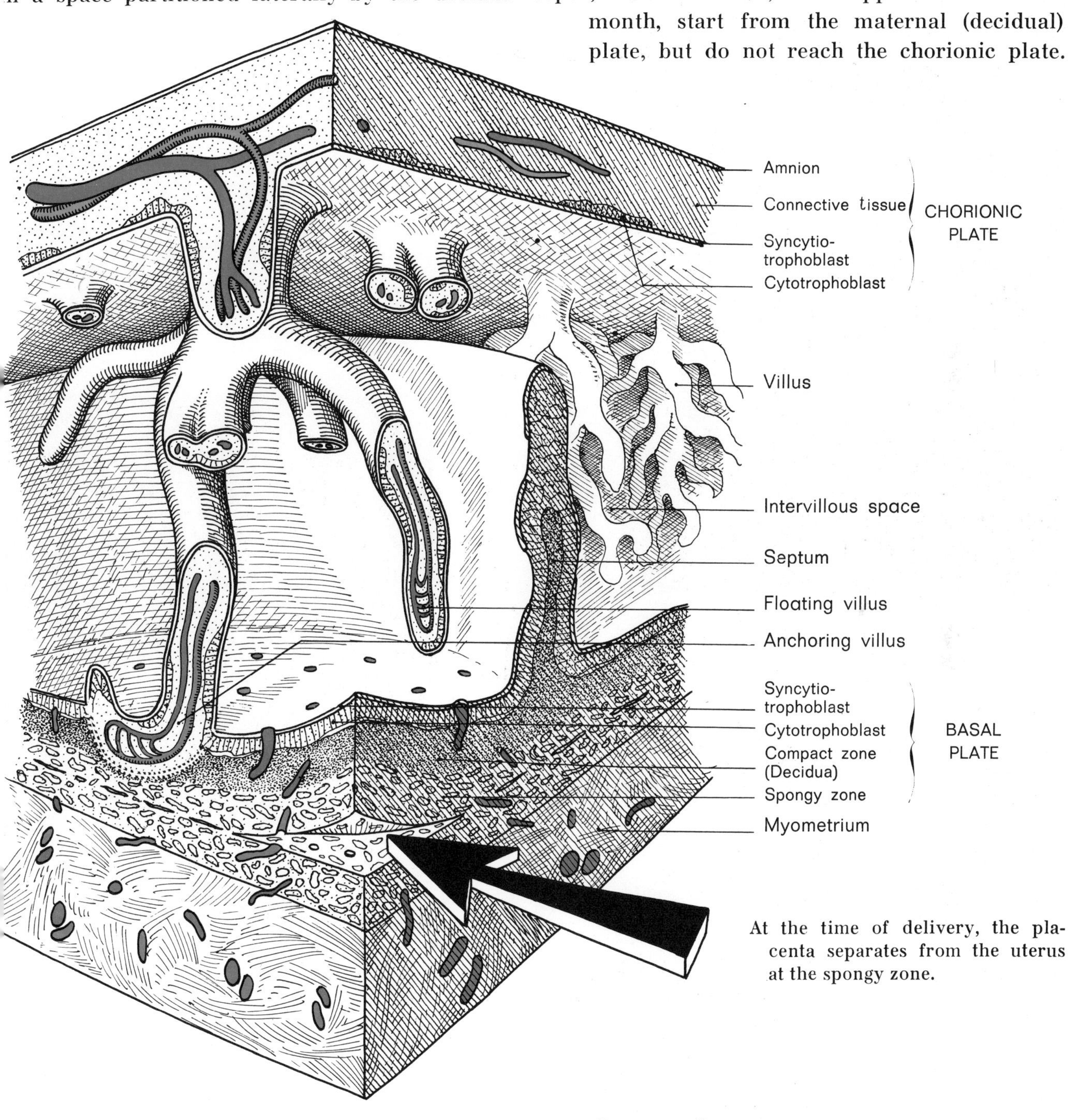

At the time of delivery, the placenta separates from the uterus at the spongy zone.

Fig. 4. — *Placenta at maturity. Composite diagram.*

Large persistent cytotrophoblastic areas are shown, particularly on the maternal plate. Actually, however, as term approaches, the cytotrophoblast disappears in this region as well, replaced by a fibrinoid layer.

THE PLACENTA :

I. — HUMAN PLACENTA
(fig. 1).

The human placenta is of the ***hemochorial type.*** The fetal tissue (chorion) is directly in contact with the maternal blood.

The *membrane* includes only 3 layers:—

— syncytiotrophoblast,
— connective tissue,
— vascular fetal endothelium.

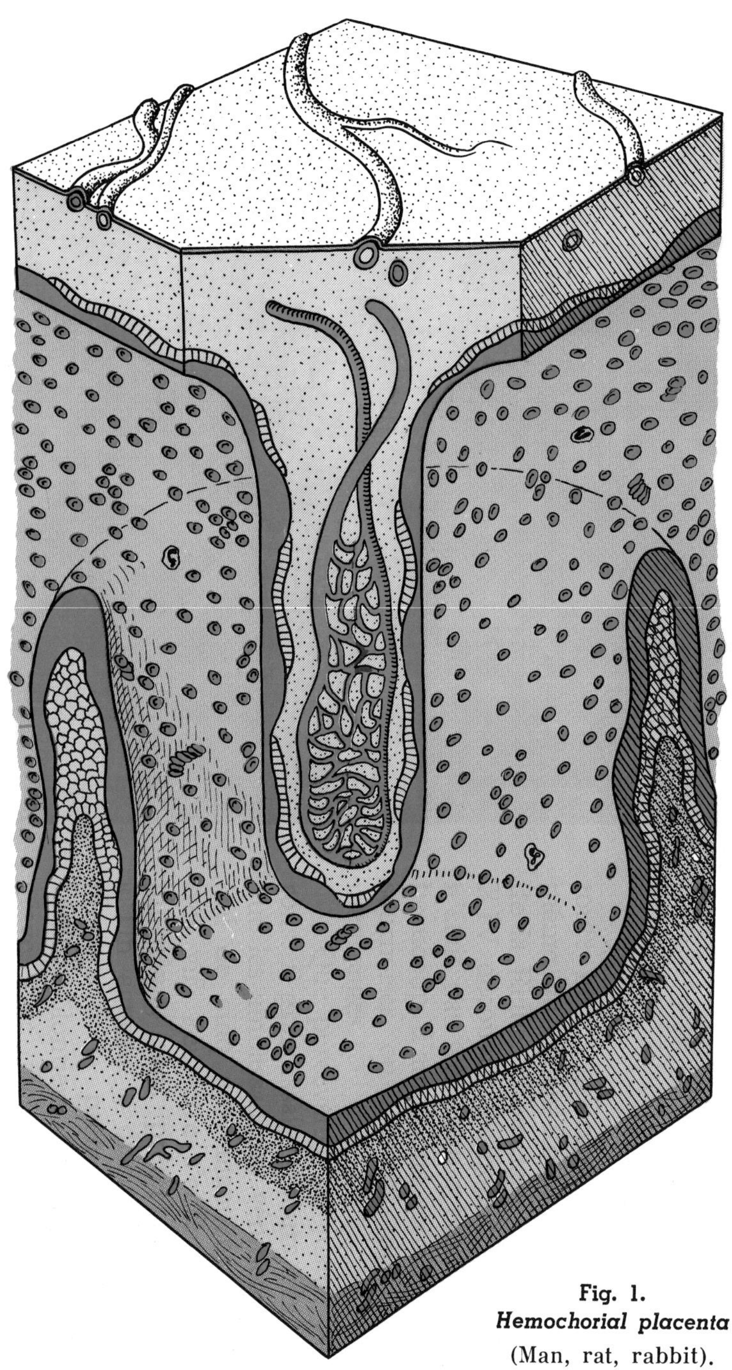

Fig. 1.
Hemochorial placenta
(Man, rat, rabbit).

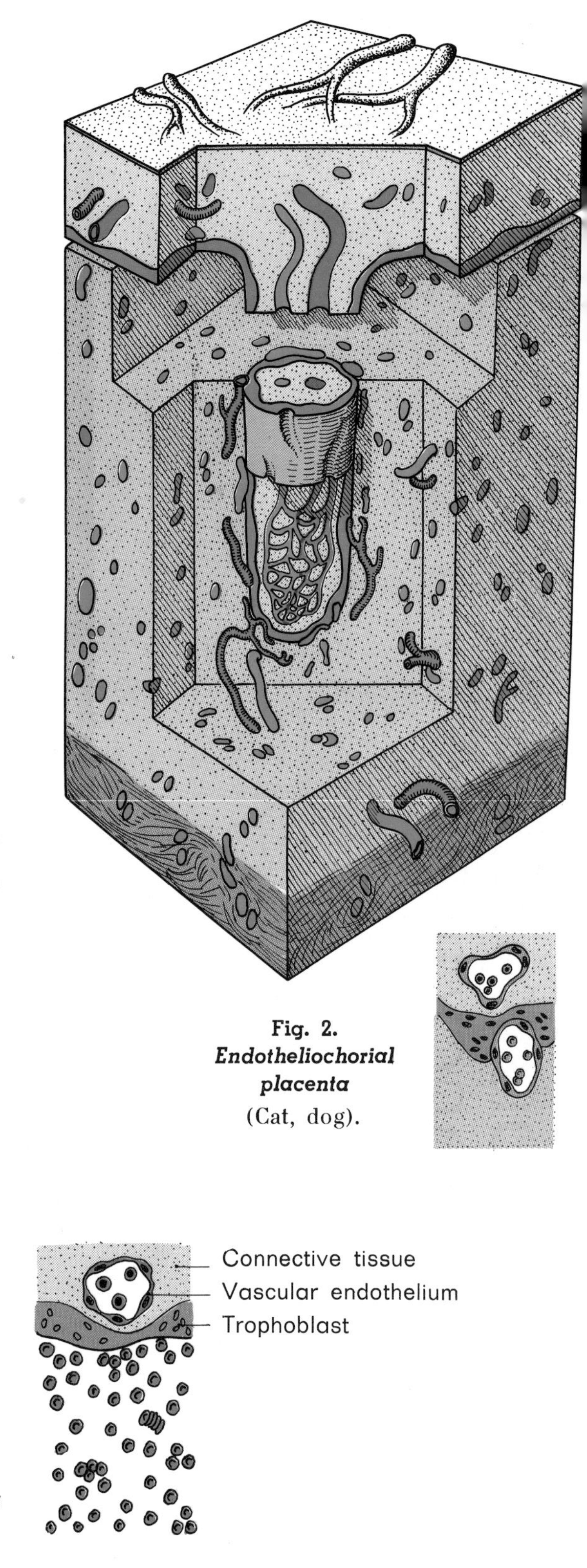

Fig. 2.
Endotheliochorial placenta
(Cat, dog).

II. — OTHER TYPES OF PLACENTAS

(fig. 2, 3, and 4.)

The membrane is much thicker.

	2	3	4	1
Endothelium of a fetal vessel				
Connective tissue of villus				
Trophoblast				
Endothelium of a maternal vessel				
Connective tissue of the maternal mucosa				
Epithelium of the maternal mucosa				

1 : hemochorial placenta; **2** : endotheliochorial placenta; **3** : syndesmochorial placenta; **4** : epitheliochorial placenta.

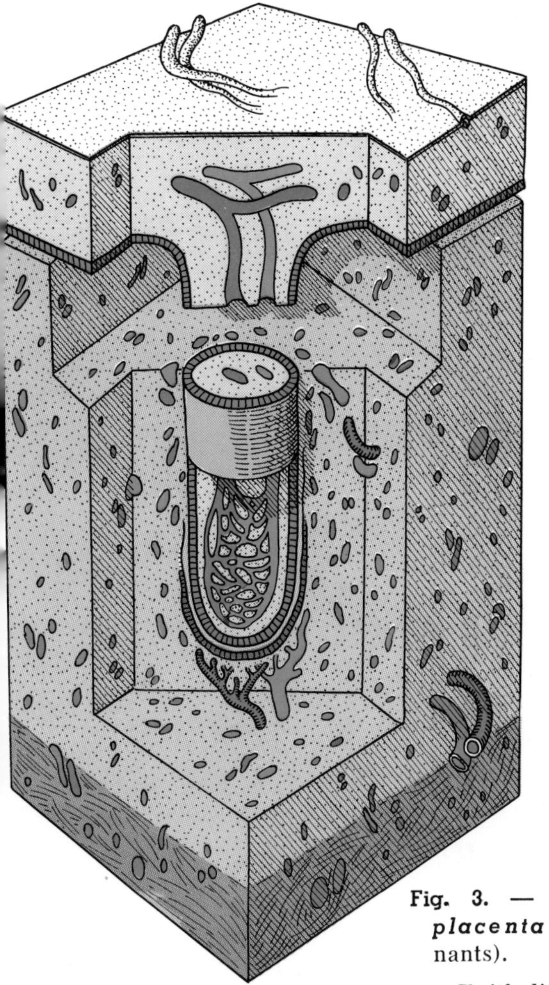

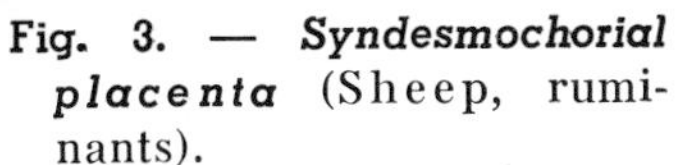

Fig. 3. — ***Syndesmochorial placenta*** (Sheep, ruminants).

Epithelium of the maternal mucosa persists partially.

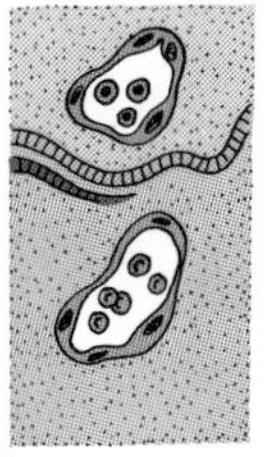

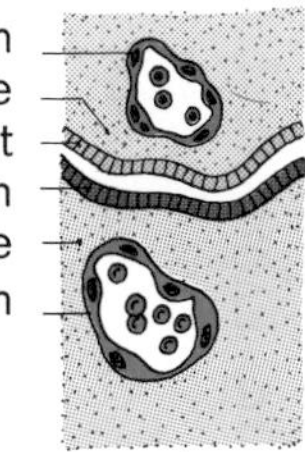

Fig. 4.
Epitheliochorial placenta
(Pig, horse).

The maternal epithelium persists, the trophoblast is simply apposed.

III. — PERMEABILITY ACCORDING

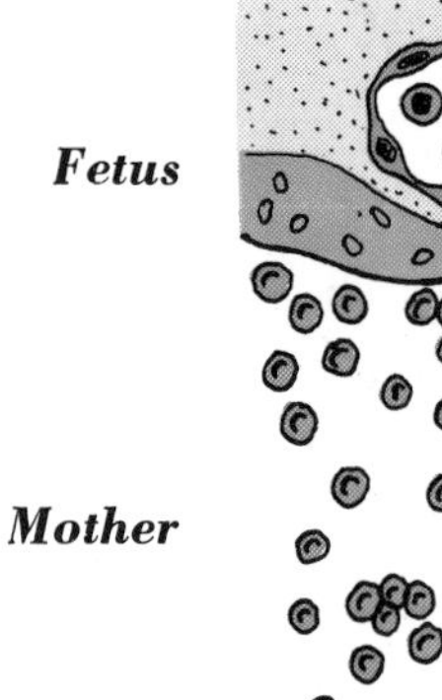

Fig. 1.
Hemochorial placenta.

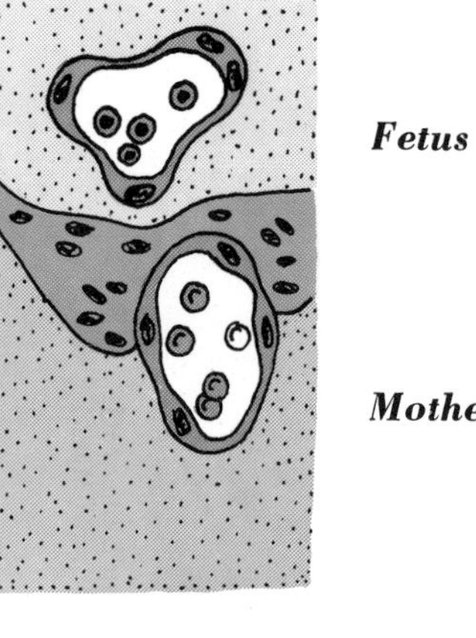

Fig. 2.
Endotheliochorial placenta.

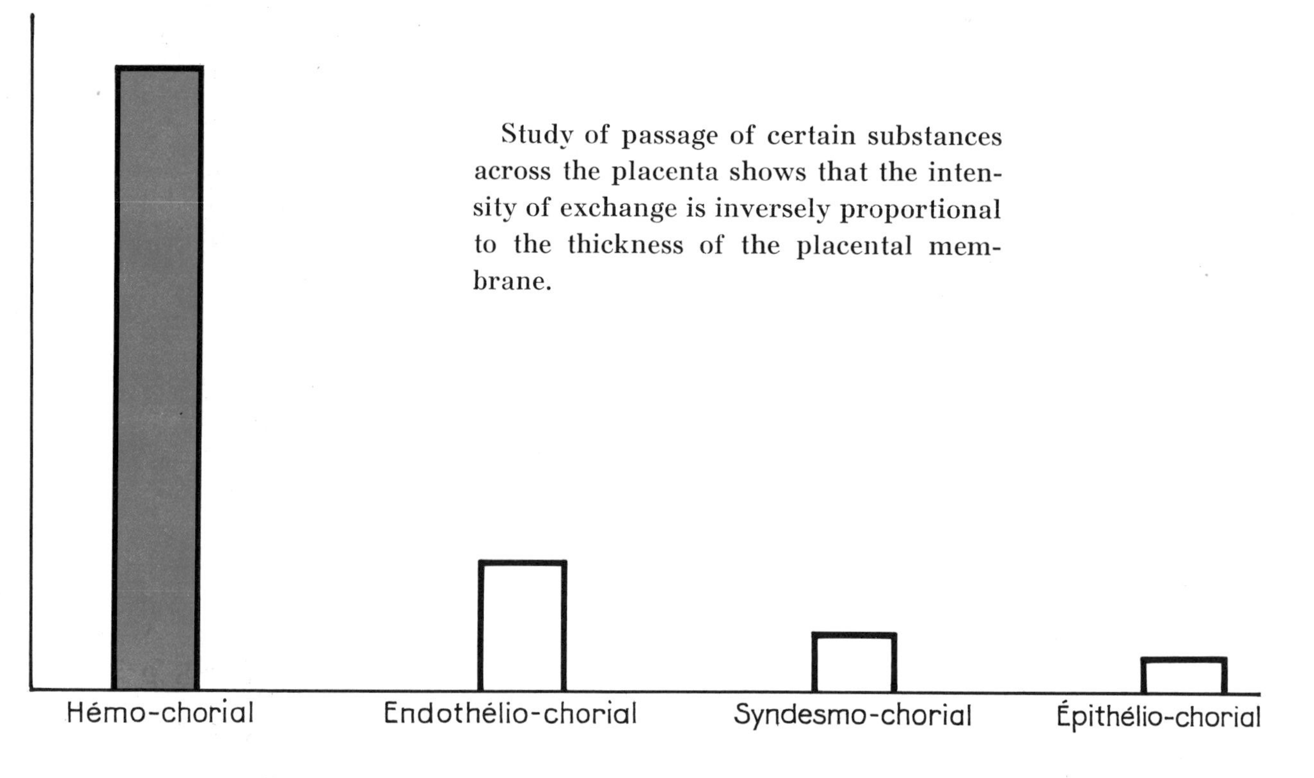

Study of passage of certain substances across the placenta shows that the intensity of exchange is inversely proportional to the thickness of the placental membrane.

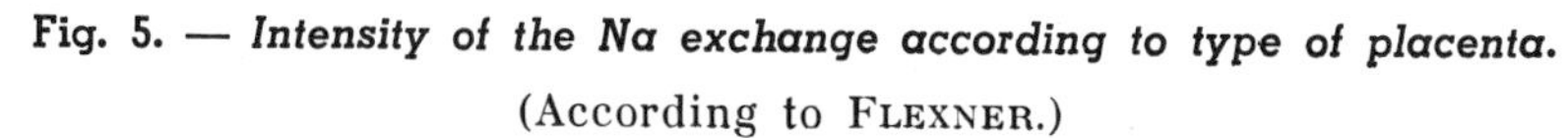

Fig. 5. — *Intensity of the Na exchange according to type of placenta.*
(According to FLEXNER.)

O TYPE OF PLACENTA

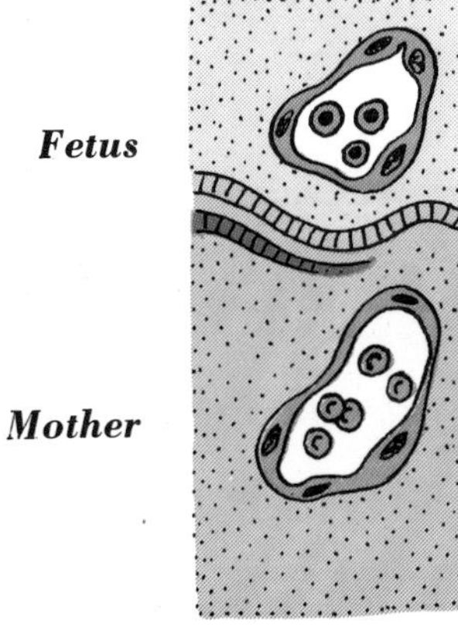

Fig. 3.
Syndesmochorial placenta.

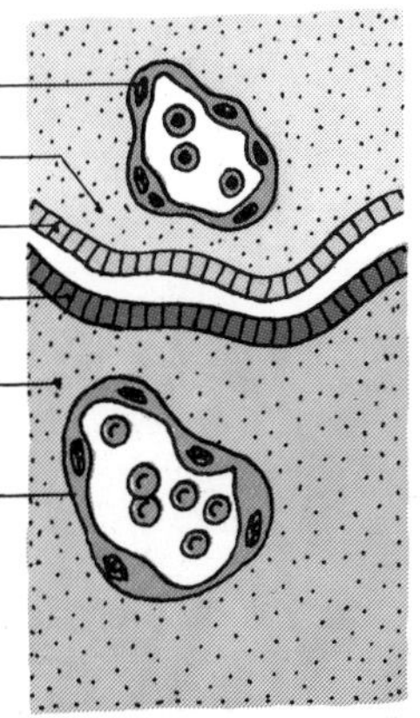

Fig. 4.
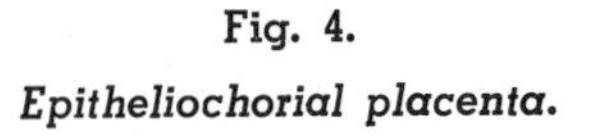
Epitheliochorial placenta.

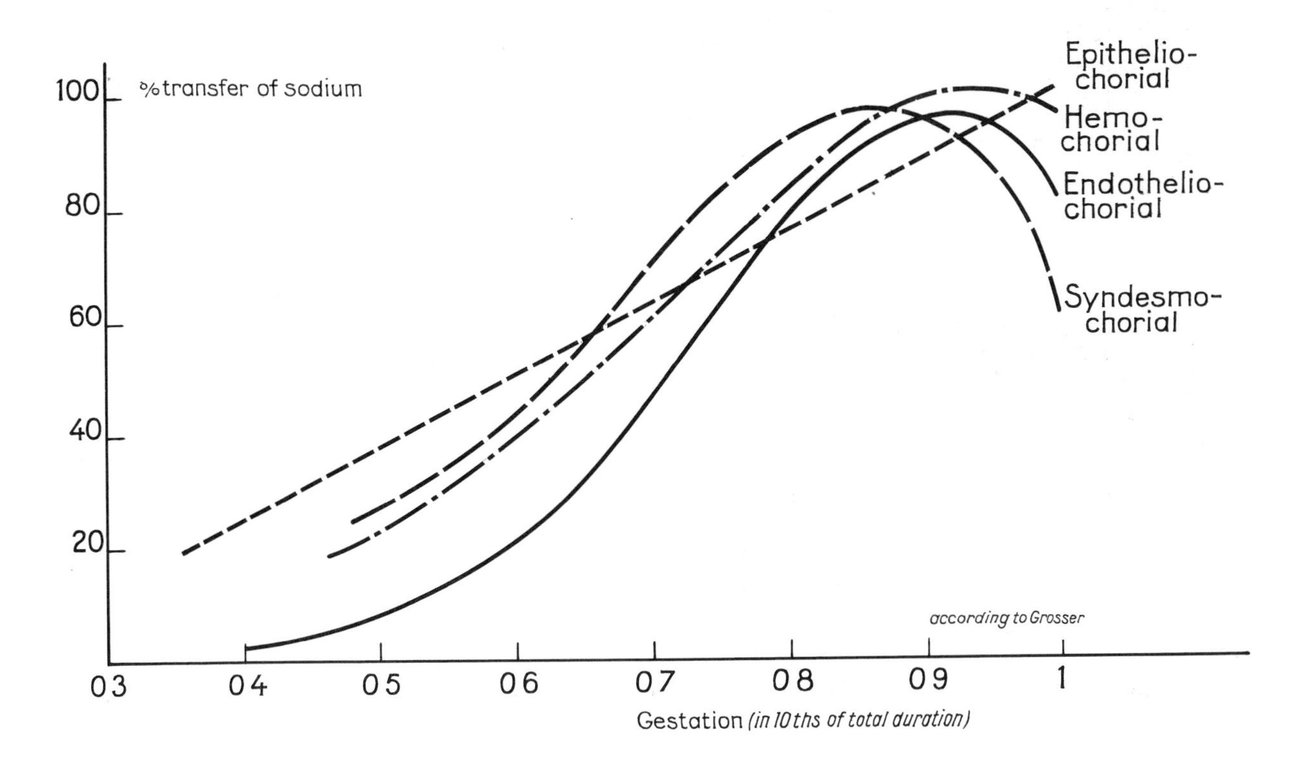

Fig. 6. — ***Development of Na exchange during gestation.***

The intensity of exchange increases regularly during gestation. In most types of placentas, it reaches a maximum shortly before normal term. The decrease at the end of gestation can be attributed, in the hemochorial placenta, to deposit of a layer of fibrinoid on the exchanging surface.

IV. — PLACENTAL MEMBRANE

The placental membrane becomes progressively thinner during pregnancy.

By the fourth month, it has a structure favorable for exchange: at this time it consists of three layers :—

— *syncytiotrophoblast;*

— *fetal vascular endothelium;*

— separated by a thin sheet of *connective tissue.*

At the end of pregnancy its thickness varies between two and six microns.

Exchange occurs both by passive diffusion and, especially, by selective and active transport resulting from activity of the membrane itself.

The exchanging surface is still further enlarged by the presence of microvilli as has been demonstrated by electron microscopy.

The brush border visible in the optical microscope (fig. 2) corresponds to these microvilli.

The electron microscope has also demonstrated an abundance of mitochondria, ribosomes, pinocytotic vacuoles and lipid enclosures.

This combination attests to intense functional activity of exchange and synthesis.

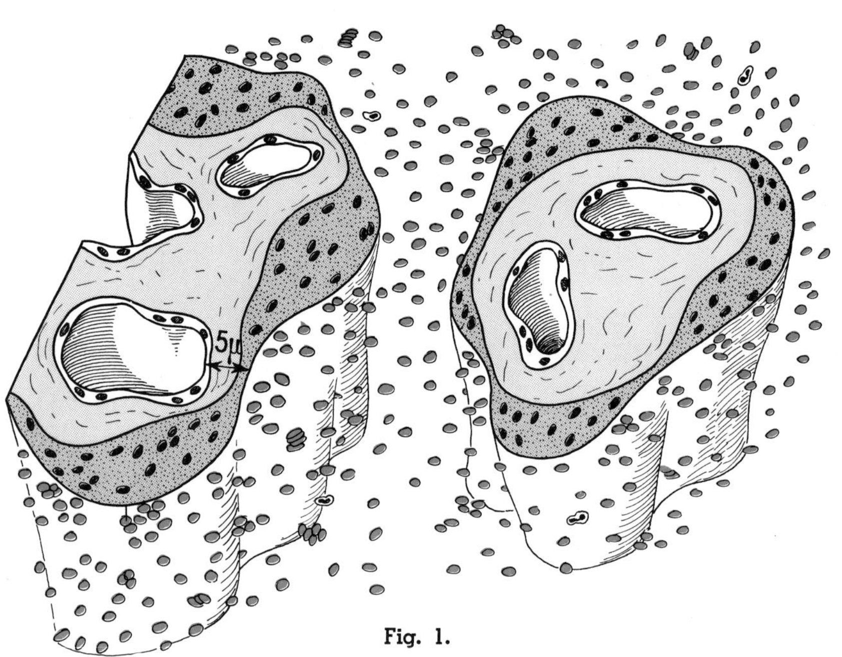

Fig. 1.

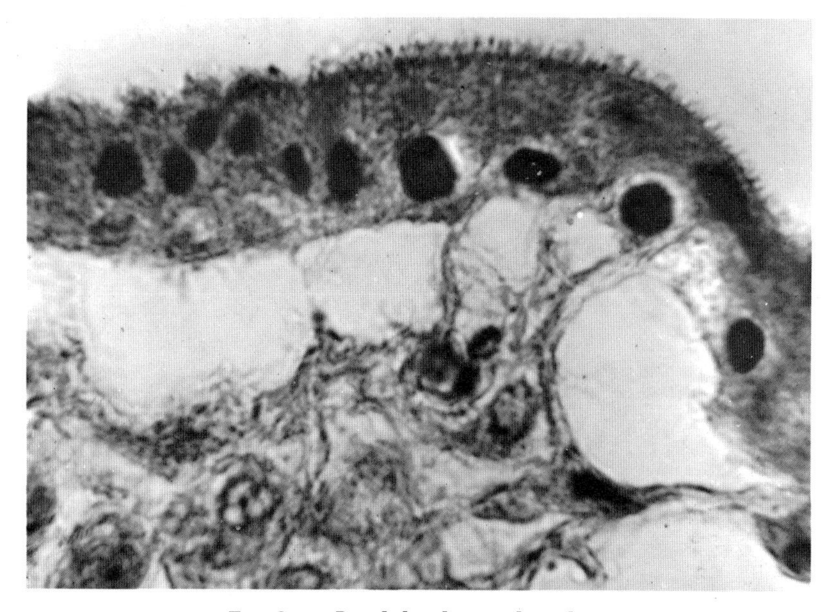

Fig. 2. — *Brush border at the edge of a placental villus* (× 1,400).

.ND FETAL-MATERNAL EXCHANGE

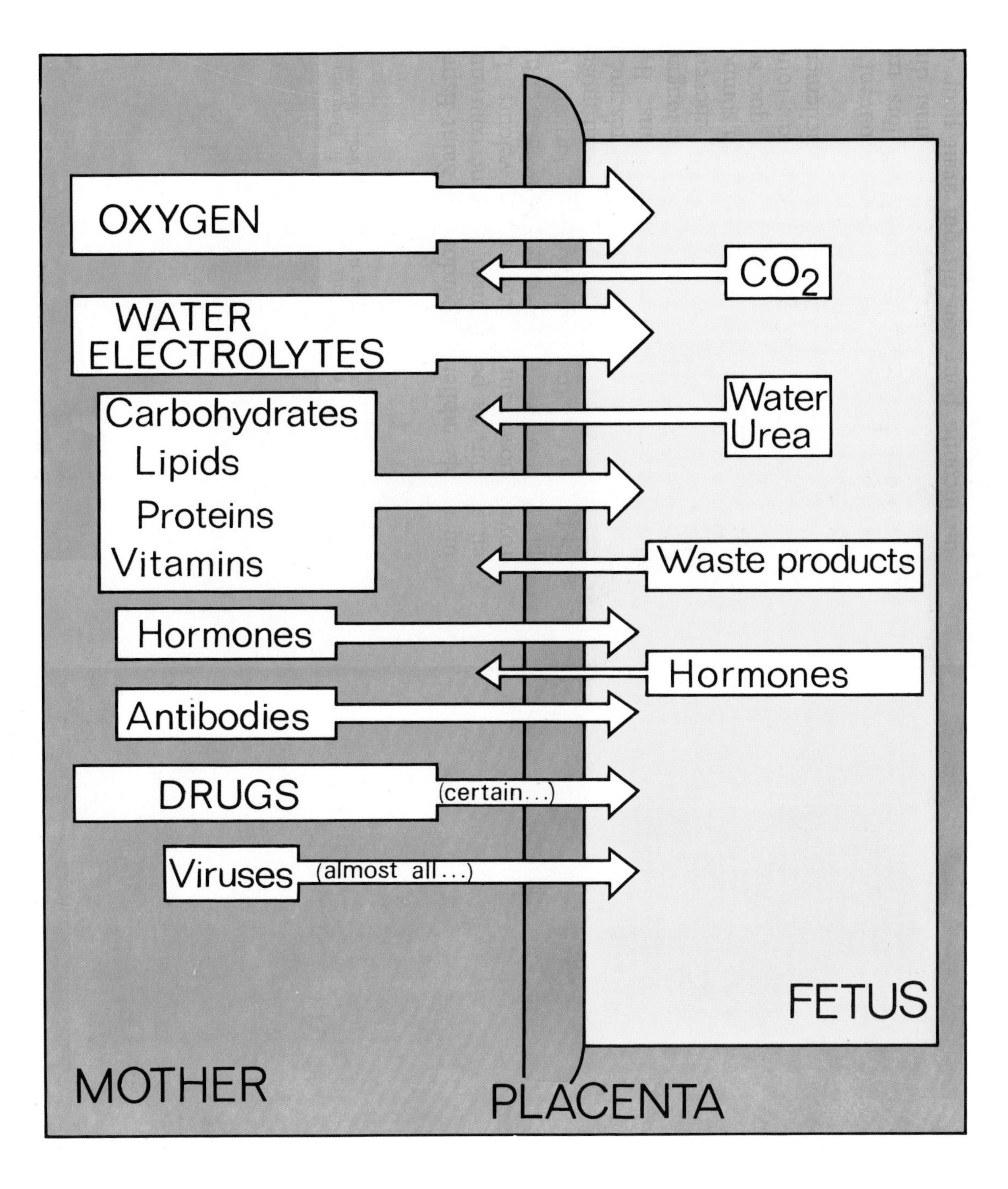

Exchange involves not only physiologic constituents, but also substances or elements which might represent a pathologic risk for the fetus.

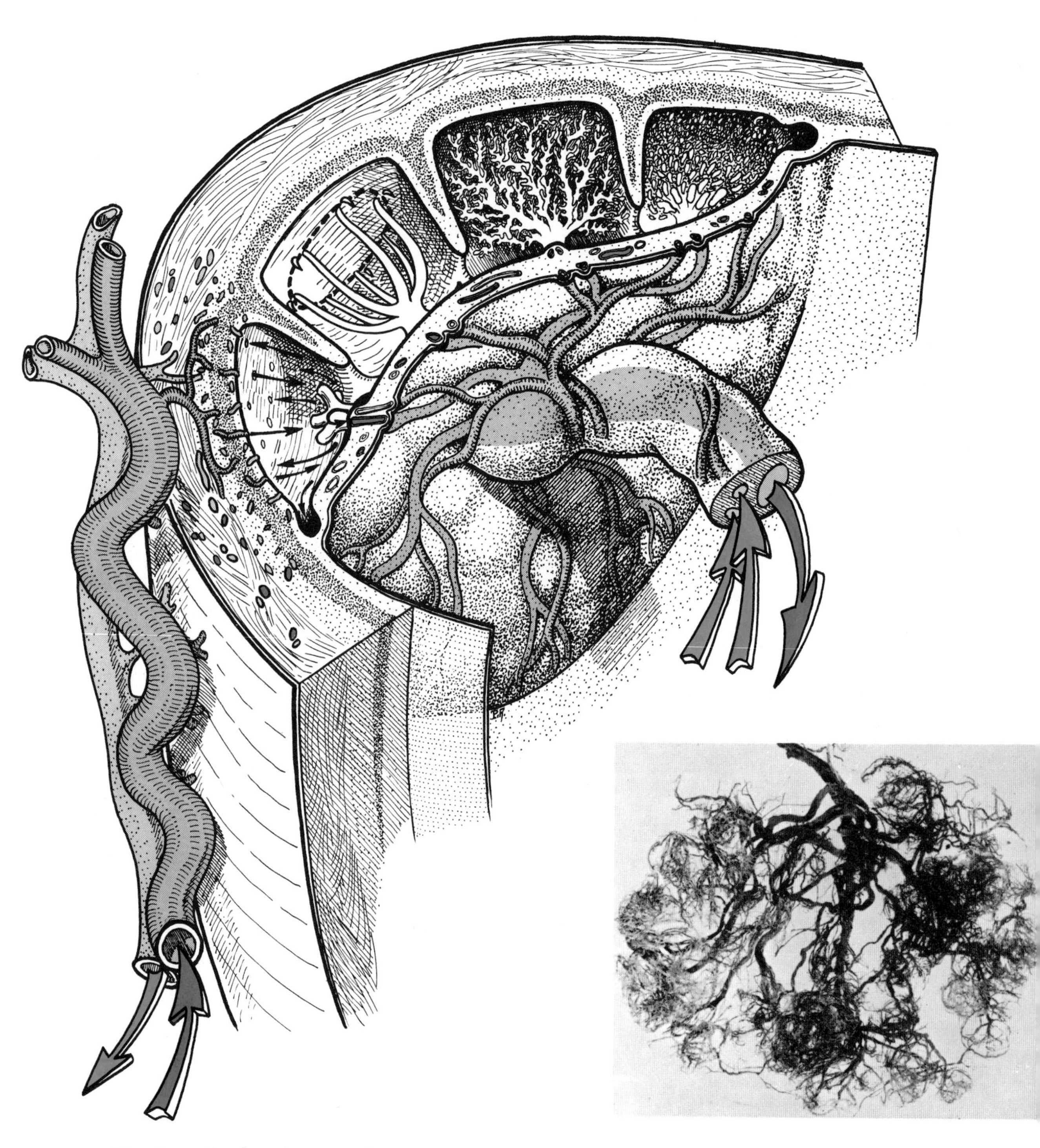

Fig. 1. — ***Fetal and maternal circulation in the human placenta.***

Fig. 2. — ***Cast of fetal vessels of a human placenta.*** Note the rich vascularity of the injected cotyledons.

CIRCULATION

The intervillous space is limited on the maternal side by the basal plate, on the fetal side by the chorionic plate, and laterally (but incompletely) by the decidual septa.

The complex division of the villi provides considerable *placental surface* : at term, it makes up more than 10 square meters. This is an important factor affecting the rate of fetal-maternal exchange.

The anchoring villi, branching from a primary trunk, are attached to the basal plate, defining a generally circular area. The villus tree in its entirety forms a complex system which constitutes the cotyledon, the anatomical unit of the placenta.

Fetal circulation : blood arrives by the two umbilical arteries, branches of the iliac arteries of the fetus. It is dispersed in an extremely dense network which penetrates the smallest villus divisions. It is taken up again by the umbilical vein and finally reaches the inferior cava system of the fetus.

This circulation is comparable to the pulmonary circulation of the adult : desaturated blood enters through the arteries, oxygenated blood returns through the vein.

Maternal circulation : blood arrives by the branches of the uterine artery. It spreads out in the intervillous spaces and circulates between the branches of the villus trees. It is taken up by branches of the uterine vein.

The flow in these two circulations is very high : about 500 ml/min. This is another factor favoring fetal-maternal exchange.

Maternal circulation results from a difference in pressure : high in the artery (70 mm/Hg), low in the intervillous space (10 mm/Hg). Blood spurts up to the chorionic plate, then comes toward the basal plate and is taken up by the uterine veins where the pressure is still lower than in the intervillous space (fig. 3).

Fetal circulation is carried out in a closed vascular system where the average pressure, estimated at 30 mm/Hg, is higher than that of the intervillous space (10 mm/Hg). This difference in pressure prevents collapse of the villus vessels.

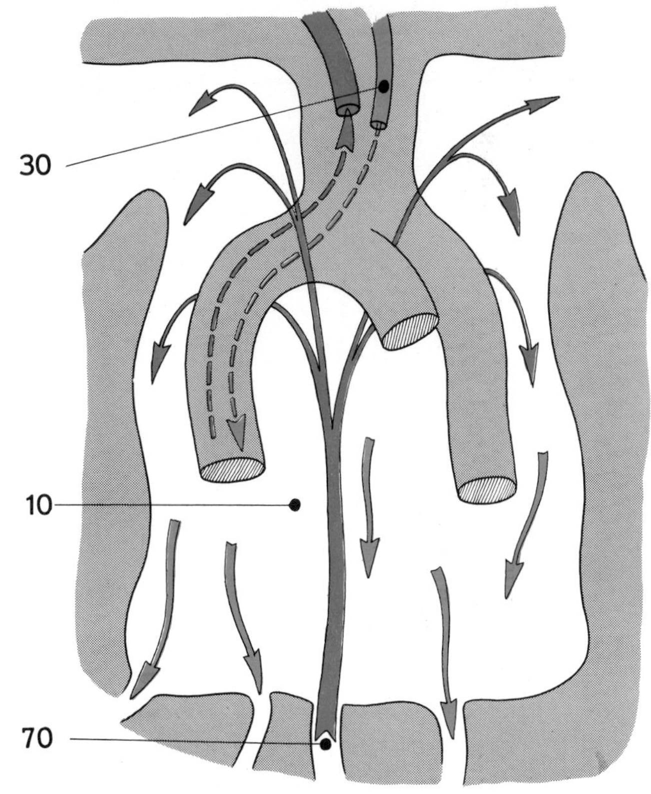

Fig. 3. — *Placental hemodynamics.*
The numbers show pressures in mm/Hg.

VI. — HORMONAL BALANCE IN PREGNANCY

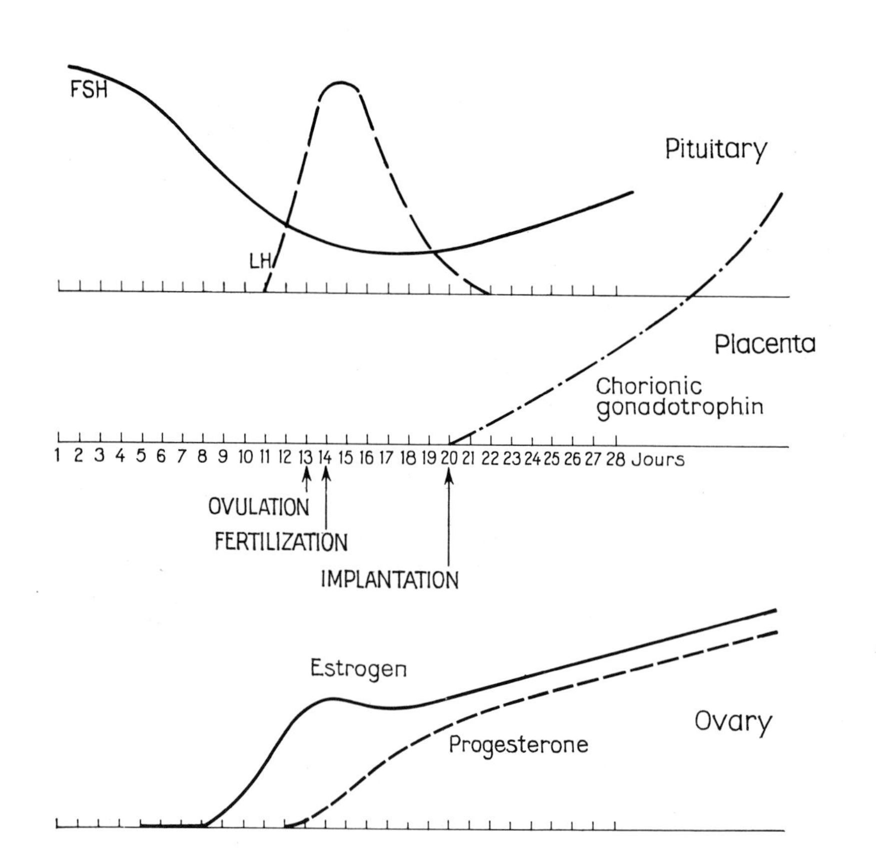

Fig. 1. — *Hormones at beginning of gestation.*

Before implantation, maintenance of pregnancy is assured by the ovarian and pituitary hormones.

After implantation, hormonal control of pregnancy is assured by the joint action of the pituitary, ovarian, and placental hormones.

Chorionic gonadotrophins are discernable very early, several days after nidation.

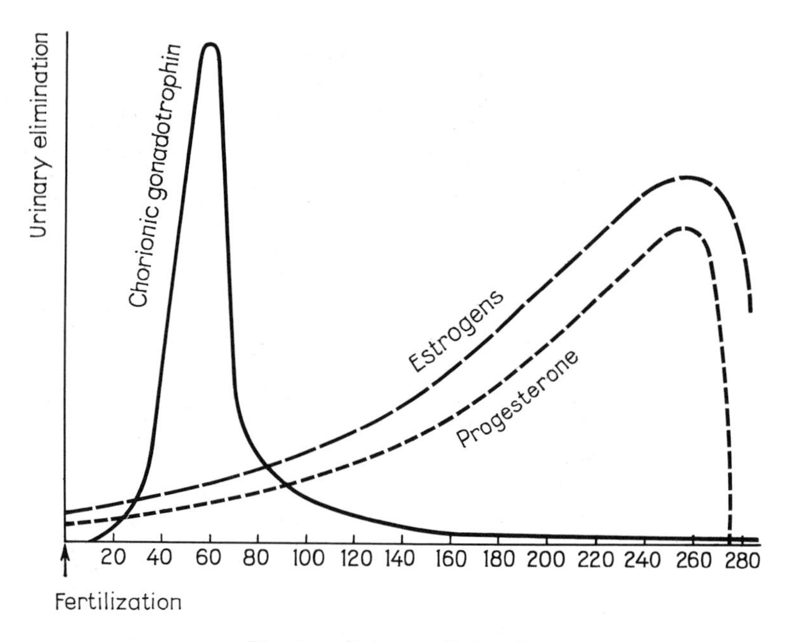

Fig. 2. — *Urinary elimination of hormones during gestation.*

Although the level of estrogens and progesterone increases regularly until term, the gonadotrophins, after reaching a peak on the 60th day, fall and are maintained at a very low level until the end of pregnancy.

VII. — PREGNANCY TESTS

Biological diagnosis of pregnancy is based on the detection of chorionic gonadotrophins. Two types of tests can be used.

1. Biological test in animals.

Biological tests can be made in various animal species : frog, toad, mouse, rat, rabbit.

The injected gonadotrophins provoke characteristic changes in the genital tract of the animal.

EXAMPLE : *biological test in the virgin female rabbit.* — Urine or blood sample from the woman being tested is injected into a virgin female rabbit that is killed 36 hours later. In a positive reaction, examination of the genital tract shows congestion and hyperemia of the uterine horns (fig. 3), and especially one or more hemorrhagic follicles in the ovaries (fig. 4).

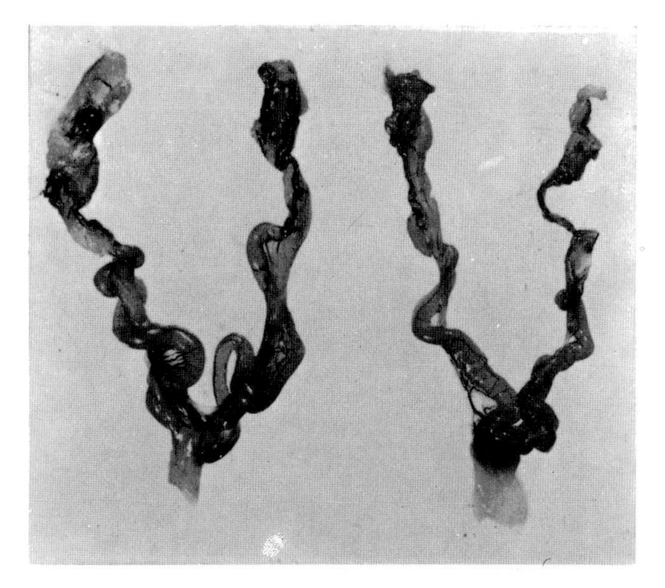

Positive reaction Control

Fig. 3. — *Biological test in virgin female rabbit.*
Entire genital tract.

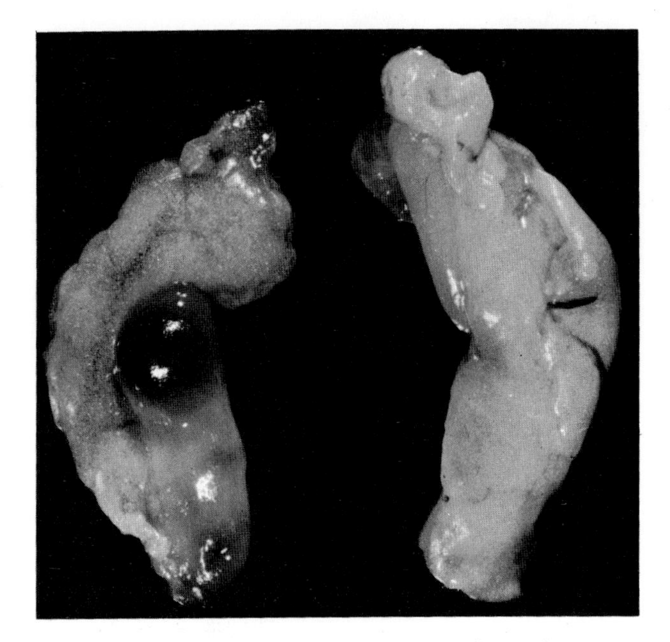

Positive reaction Control

Fig. 4.
Biological test in virgin female rabbit.
Ovaries.

2. *Immunological test.*

The presence of chorionic gonadotrophins in the urine is shown b antigonadotrophic serum (obtained by immunization of an animal again human gonadotrophin).

The reaction system contains two components :—

— antigonadotrophic serum (reactive agent *a*, fig. 1);
— red blood cells artificially covered with gonadotrophin (reacti agent *b*, fig. 2).

Fig. 1. — *Serum antigonadotrophin* (reactive agent *a*).

If *a* and *b* are mixed, antibodies in the serum agglutinate the red bloc cells through the gonadotrophins which are attached : this phenomenon directly visible in a test tube or on a slide. This is the control reactic (fig. 3).

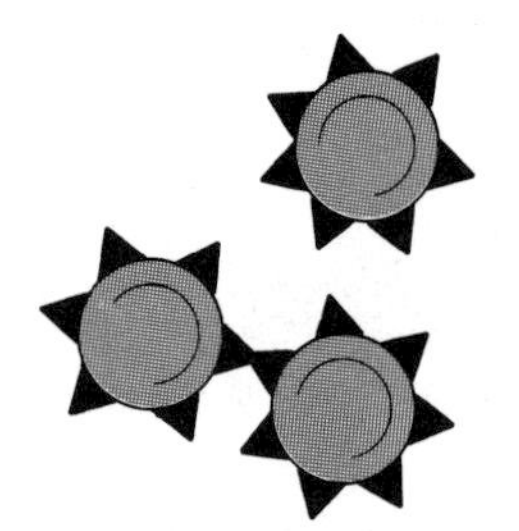

Fig. 2. — *Red blood cells covered with gonadotrophin* (reactive agent *b*).

If, first, several cc of urine containing free gonadotrophins (from a pregnant woman) are mixed with reactive agent *a*, these gonadotrophins block the serum *a* antibodies and prevent them from agglutinating the red blood cells *b* (fig. 4).

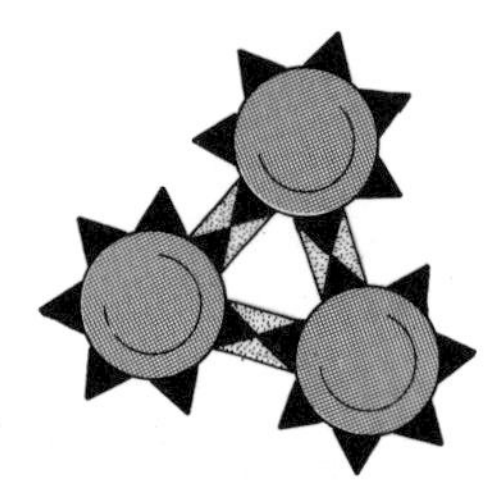

Fig. 3.
Agglutination : negative reaction.

Fig. 4. — *No agglutination : positive reaction.*

Thus :—

NON-PREGNANT WOMAN	PREGNANT WOMAN
1. Urine + a. 2. Urine + a + b. ↓ *Agglutination.*	1. Urine + a. 2. Urine + a + b. ↓ *No agglutination.*
Negative reaction.	**Positive reaction.**

Currently, the red blood cells of reactive agent *b* are replaced by inert particles (latex, for example) which play the same role and have the advantage of not changing with time.

VIII. — HYDATIDIFORM MOLE

Hormonal equilibrium is reliable evidence of the progress of pregnancy : it makes possible surveillance of pregnancy as well as its diagnosis.

Hormonal levels also permit diagnosis of certain anomalies, particularly placental degeneration, such as *hydatidiform mole* (fig. 5, 6 and 7), and its malignant transformation, *chorioepithelioma*. In these cases, chorionic gonadotrophins are abnormally elevated.

Fig. 5. — ***Macroscopic appearance of molar placenta.***

Fig. 6. — ***Detail of molar villi in grape seed form.***

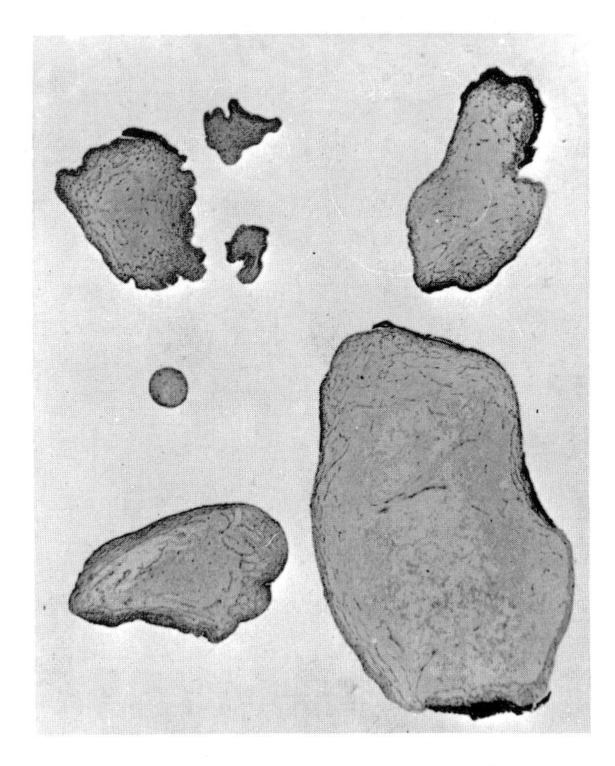

Fig. 7. — ***Microscopic section of molar villi*** **:** the core of the villi is uniquely formed from edematous connective tissue without vascularization. Such a placenta is incapable of assuring survival of the fetus.

Twins occur in about one percent of all pregnancies.
There are two types of twins.

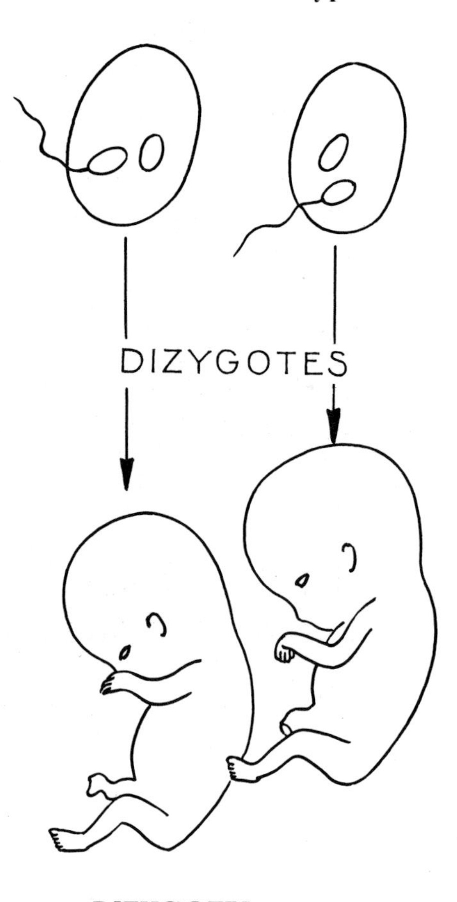

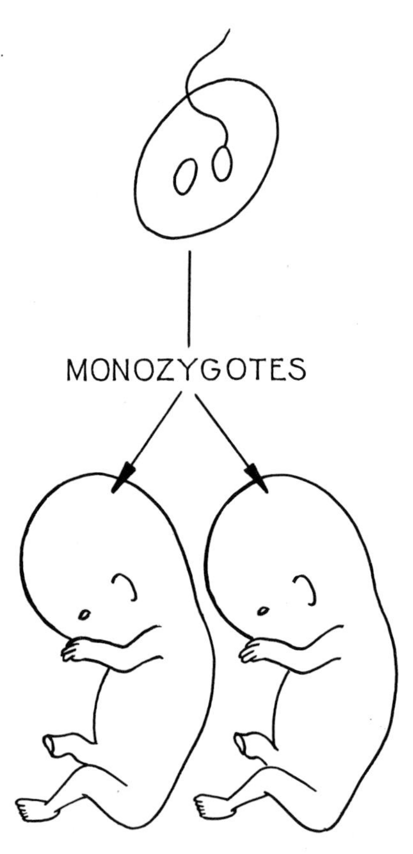

Fig. 1.

DIZYGOTES

or diovular,
or divitelline.

— About 70 % of twins,
— and 7 in 1,000 births.

Developing from two different ova, these children have the following characteristics :—

— physical dissimilarity,
— possible difference in sex,
— difference in blood characteristics,
— no tolerance for transplants.

Theoretically, the fetal membranes should be separate; in fact, however, the placentas may be confluent and the membranes partially joined.

MONOZYGOTES

or monovular,
or monovitelline.

— About 30 % of twins,
— and 3 in 1,000 births.

Developing from a single egg, they have the following characteristics :—

— morphologically and physiologically identical,
— same sex,
— identical blood characteristics,
— tolerance for transplants.

The fetal membranes are shared more or less completely according to the *stage of separation* (fig. 3).

In both cases there is hereditary predisposition : there are families where twin pregnancies are especially frequent.

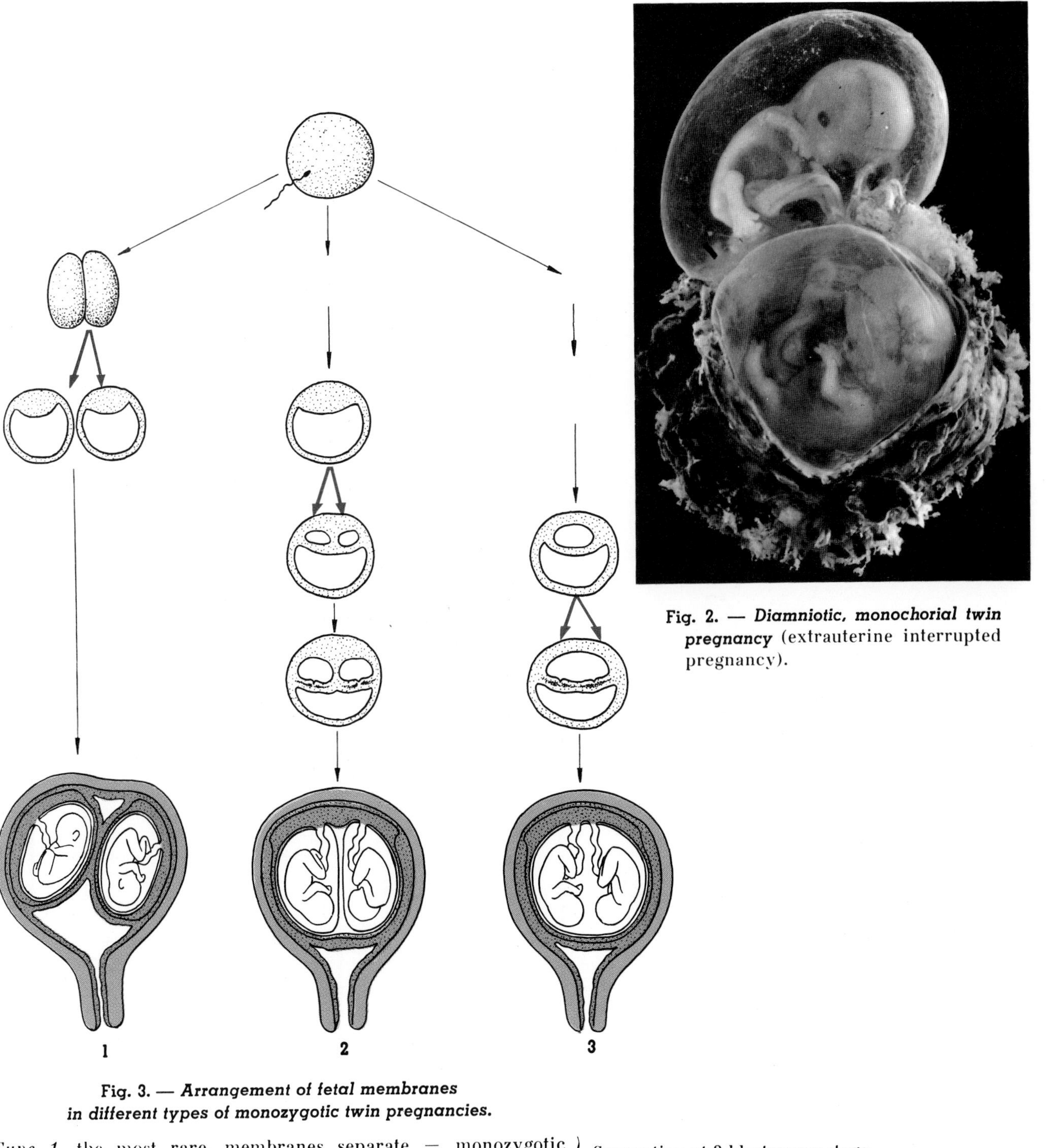

Fig. 2. — ***Diamniotic, monochorial twin pregnancy*** (extrauterine interrupted pregnancy).

Fig. 3. — ***Arrangement of fetal membranes in different types of monozygotic twin pregnancies.***

Type 1, the most rare, membranes separate = monozygotic dichorionic twin pregnancy.	Separation at 2 blastomere stage.
Type 2, the most frequent : common placenta, separate amnion = monozygotic, monochorionic, diamniotic twin pregnancy.	Separation at embryoblast (inner cell mass) stage.
Type 3, completely common membranes = monoamniotic twin pregnancy.	Separation at embryonic disc stage.

PRINCIPAL STAGES

I. — EMBRYONIC

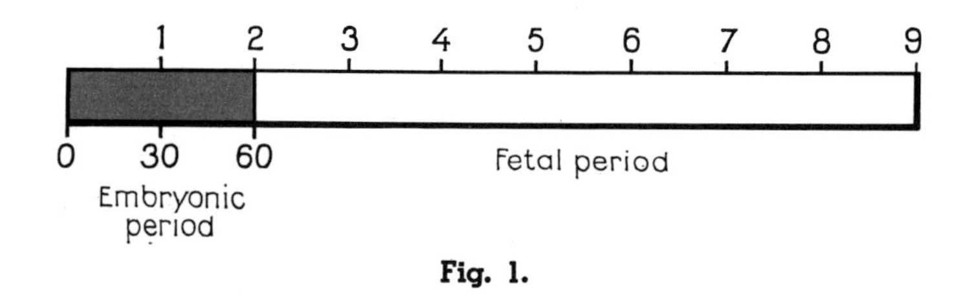

Fig. 1.

Despite its short duration, the embryonic period is of fundamental importance, as the embryo acquires its almost definitive form *(morphogenesis)* and builds the main outlines of its organs *(organogenesis)*. During the long fetal period which follows, the organs undergo little more than maturation at the histological level *(histogenesis)*.

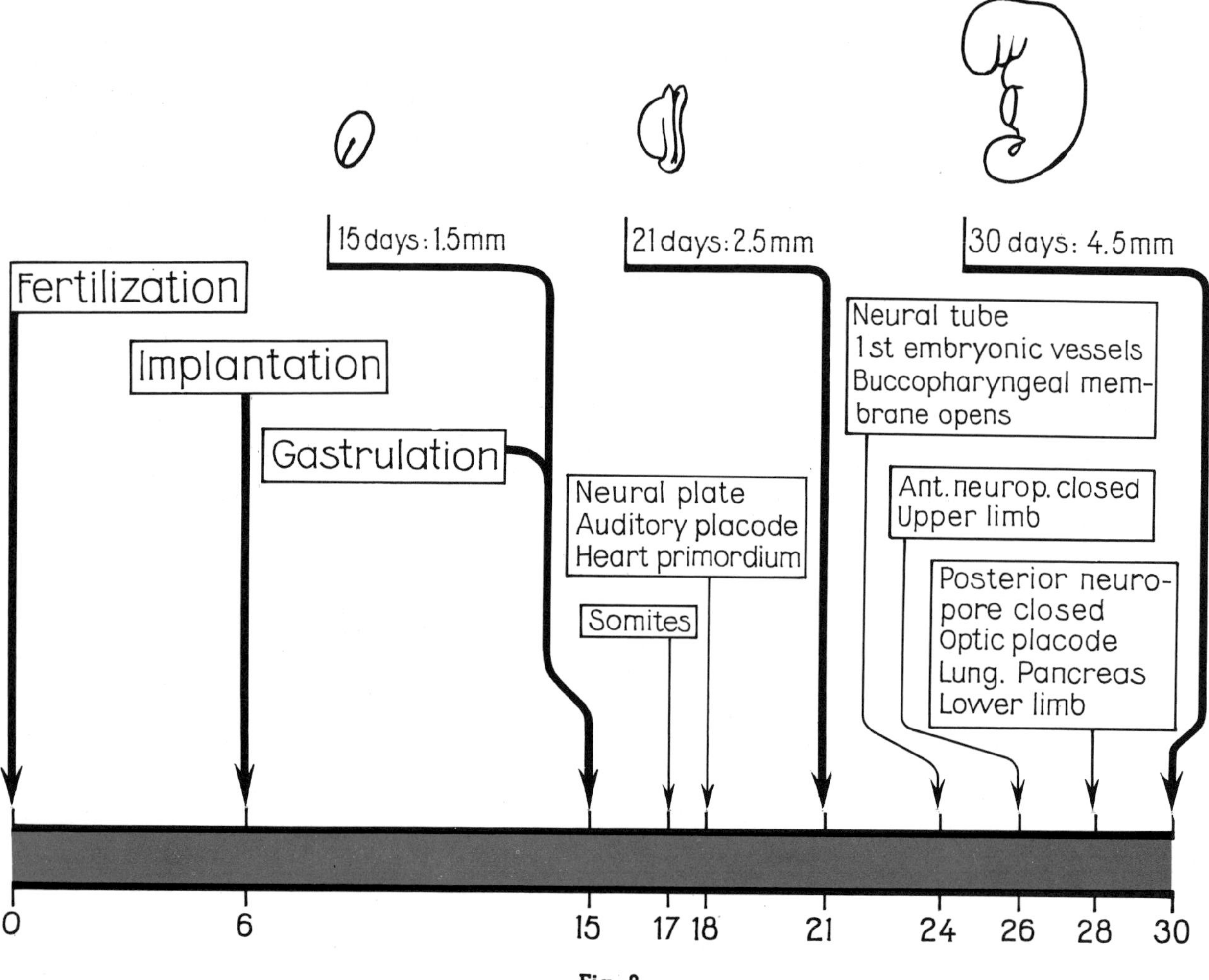

Fig. 2.

)F DEVELOPMENT

ERIOD

The dates of appearance of primordia shown
ı figures 2 and 3 correspond to the data given
y a majority of authors.

The figures are on a scale of 5 to 1.

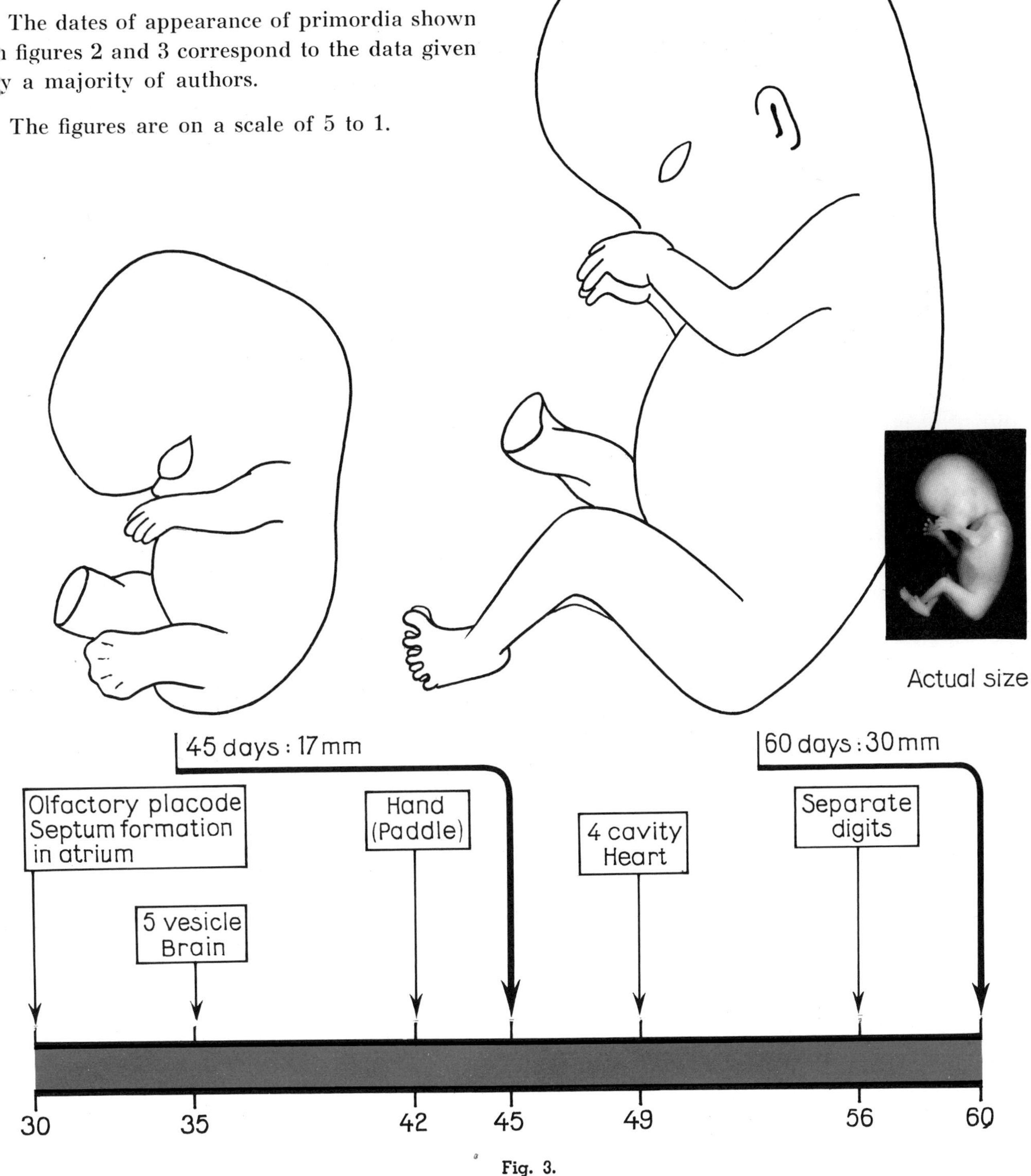

Fig. 3.

II. — FETAL

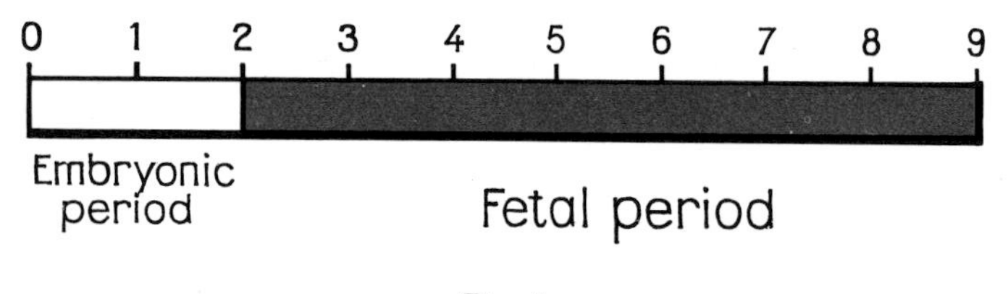

Fig. 1.

By the second month, the embryo has all the basic outlines of its organs and its appearance is already clearly defined : from now on it is a fetus (fig. 2).

During the following 7 months, the fetal organism accomplishes maturation of its primordia, reorganizes their spatial relationships, and begins to make functional use of its organs for part of its needs. Finally, during this period, it grows considerably : its size goes from 30 to 330 mm (see fig. 3), and its volume increases proportionally.

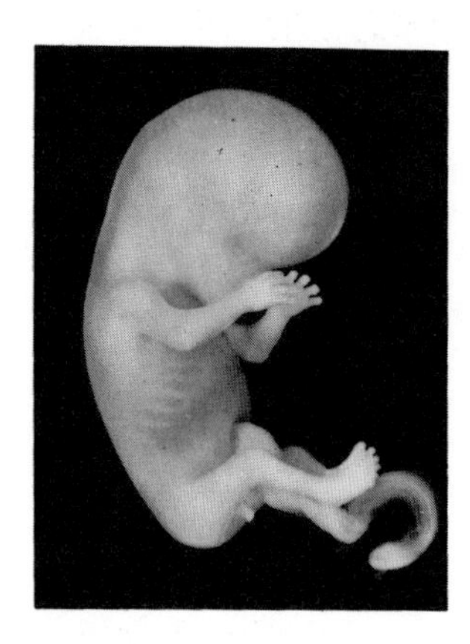

Fig. 2. — *Human fetus, 4 cm (70 days).* Actual size.

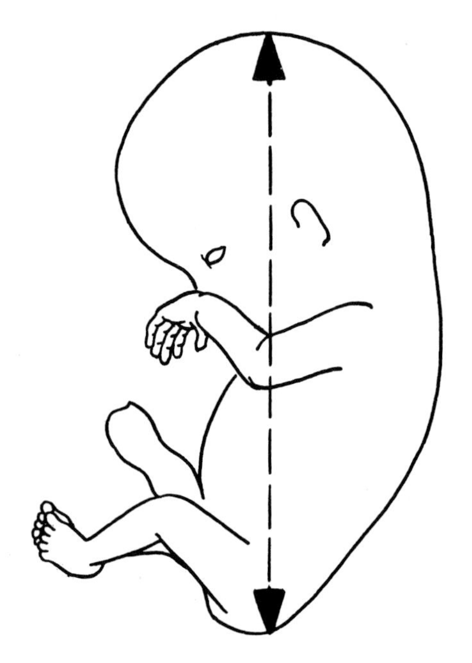

Fig 3. — *Average size of embryo and fetus in crown-rump length :—*

2 weeks	1.5 mm
3 weeks	2.5 mm
4 weeks	5 mm
5 weeks	8.5 mm
7 weeks	20 mm
2 months	33 mm
3 months	95 mm
4 months	135 mm
6 months	230 mm
9 months	335 mm

(500 mm for crown-heel length).

PERIOD

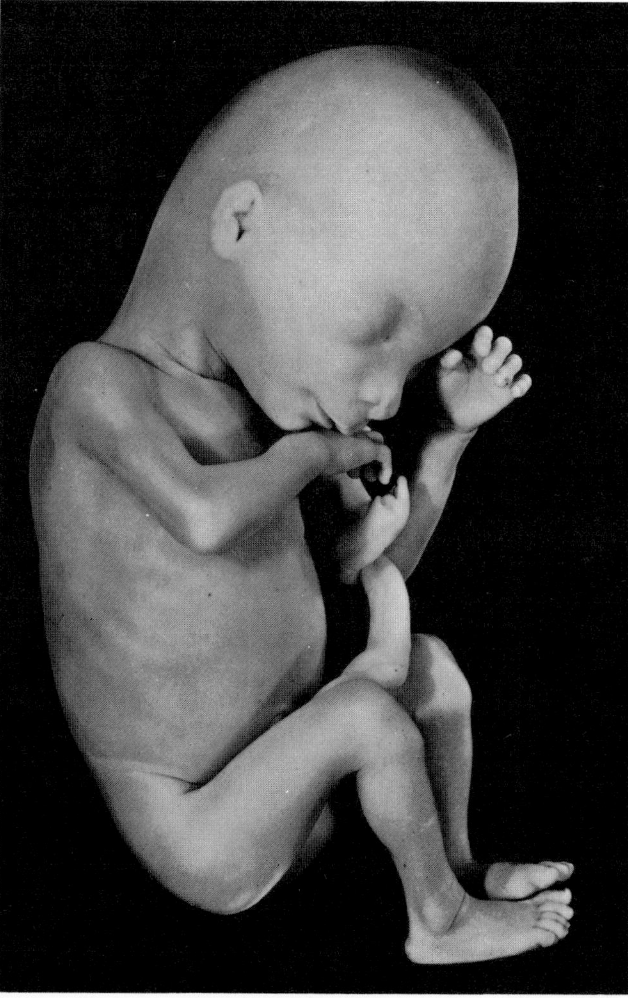

Fig. 4. — ***Human fetus, 12 cm (3 1/2 months).*** Actual size.

Fetal growth is a much more complex phenomenon than simple calculations of size and volume would lead one to suppose. It results, in fact, from the sum of very *asynchronous* growth of different parts and organs: the body proportions of a fetus at term are very different from those of a fetus at 2 months (fig. 5).

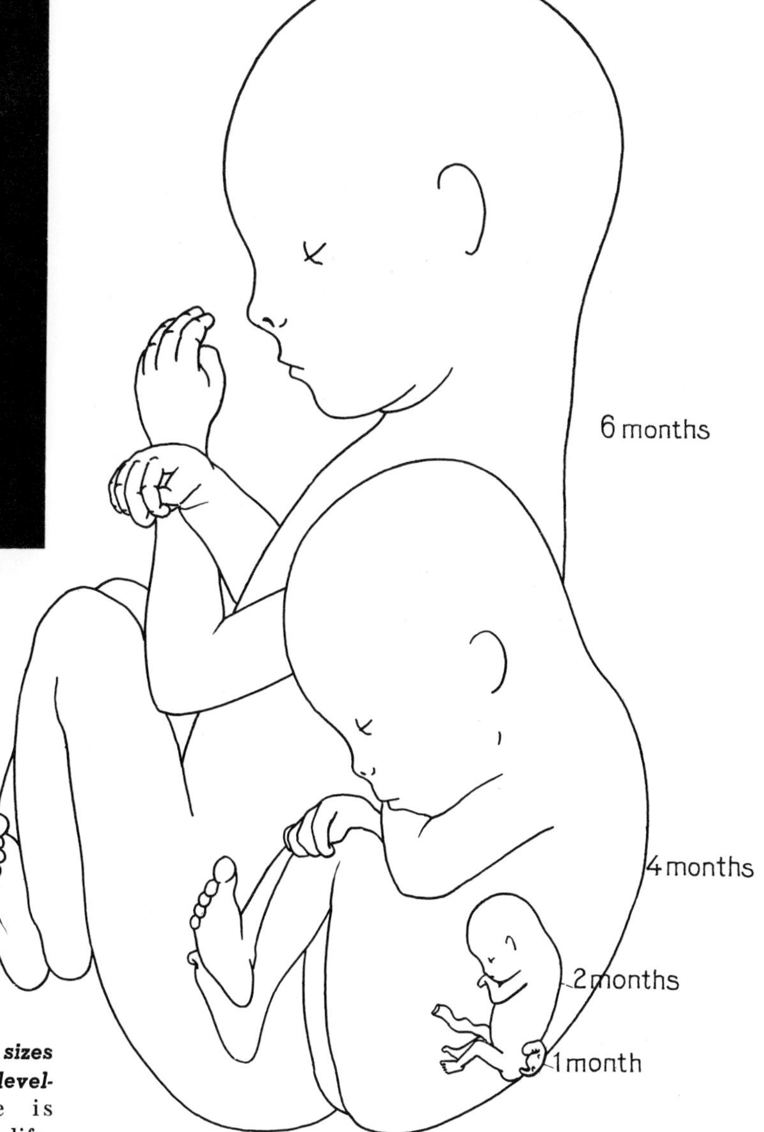

Fig. 5. — ***Comparative sizes at different stages of development.*** The scale is slightly smaller than life.

GENERAL MECHANISMS

In order to produce a complete organism beginning with the zygote, normal development involves *growth* and *differentiation,* both under strict coordination and rigorous organization.

The processes which accomplish this harmonious development depend upon two overall controls :—

— one is ***genetic*** : it determines the inherent potential of the organism;
— the other is called ***epigenetic*** : it assures progressive formation of the primordia, then of the definitive organs and includes an ensemble of complex mechanisms acting simultaneously or successively :—

1. **Cellular movements.** — Gastrulation is the best example, based entirely, as we have seen, on migration and invagination of primary ectoderm cells. Another example is shown by migration of sclerotome cells toward the notochord region to furnish material for the future spinal axis (fig. 1).

2. **Induction.** — Reciprocal influences between cellular groups, which will be analyzed at greater length later. The notochord is the best known inductive tissue (fig. 2) and its role is easy to demonstrate (see p. 95).

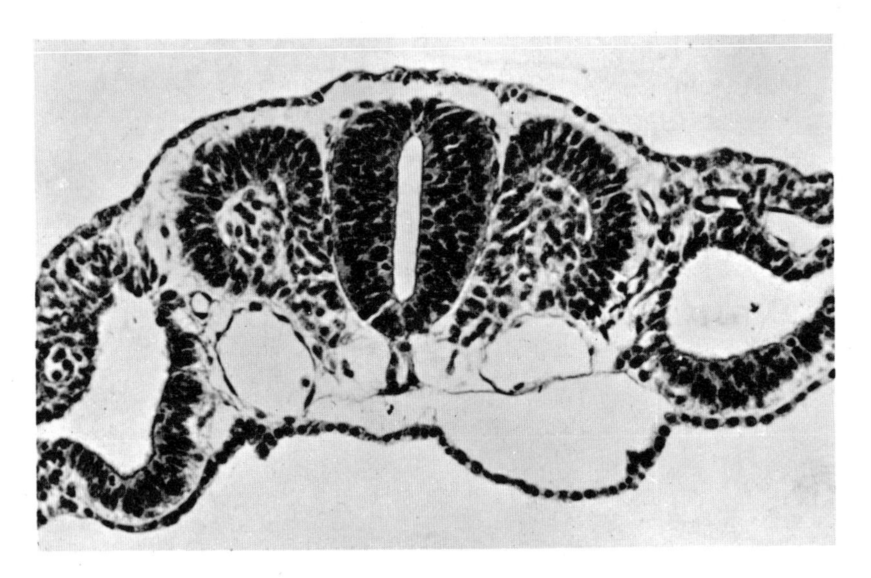

Fig. 1. — *Cellular migration :*
the sclerotome, arising from the somite, migrates toward the notochord region.

)F NORMAL DEVELOPMENT

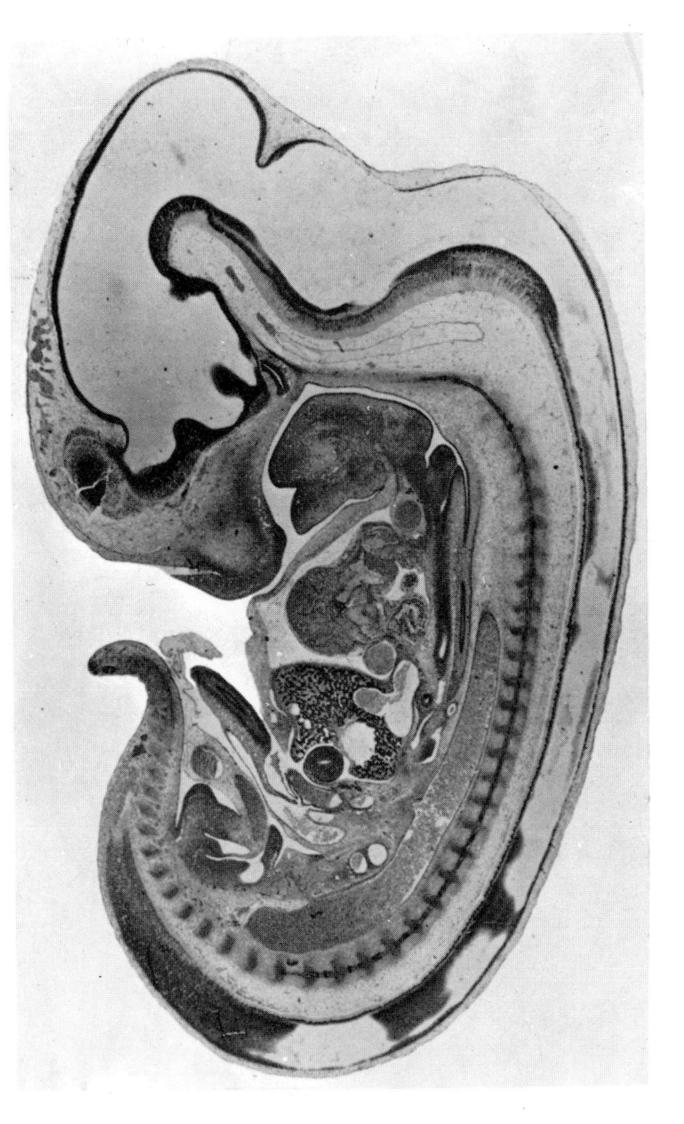

'ig. 2. — *Induction:* the notochord has acted as inductor on the sclerotome, which forms the vertebral primordia.

3. **Regression.** — The notochord is also an example of this process, since it disappears almost entirely after having contributed to formation of the vertebral bodies (fig. 3).

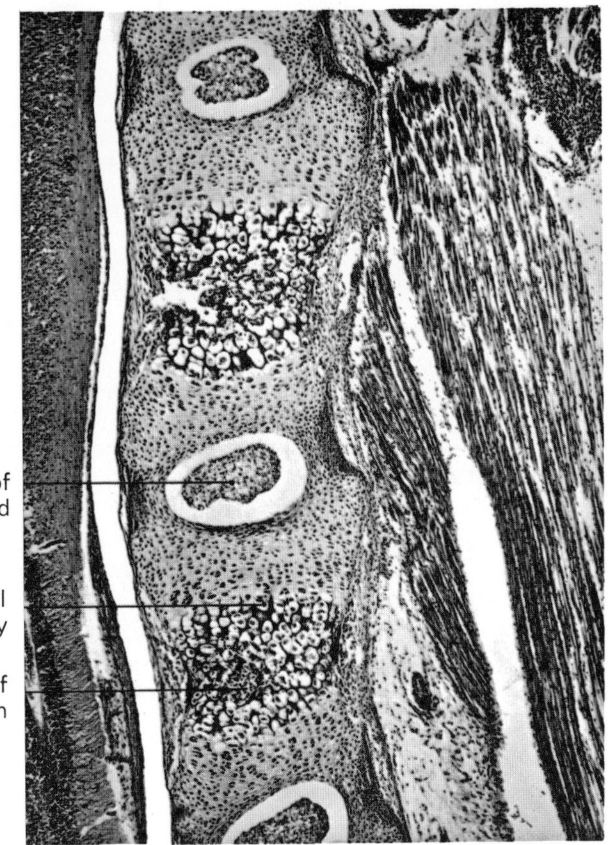

Fig. 3. — ***Regression:*** development of the vertebral bodies is accompanied by regression of the notochord, which persists further only at the level of the intervertebral discs.

4. **Regulation.** — This is the property shown by an egg of reconstituting a harmonious whole when part of its substance s removed. Monozygotic twins are the most apparent spontaneous demonstration of the existence of the phenomenon of regulation in the human species.

INDUCTIO

Induction, essential determinant of embryonic development, is the process by which a cellula group called the *inductor* provokes differentiation of another cellular group called the *competen*

The primary inductor. — Demonstrated by SPEMANN, who, experimenting on the amphibia egg, discovered the major inductive role of the dorsal lip of the blastopore (that is, the futur notochord substance) (fig. 1).

Thus, material coming from blastopore dorsal lip is shown to be capable of inducing, in tissu normally having another destiny, the differentiation of a new organism. Because of this, it i designated the *primary inductor* or organizer.

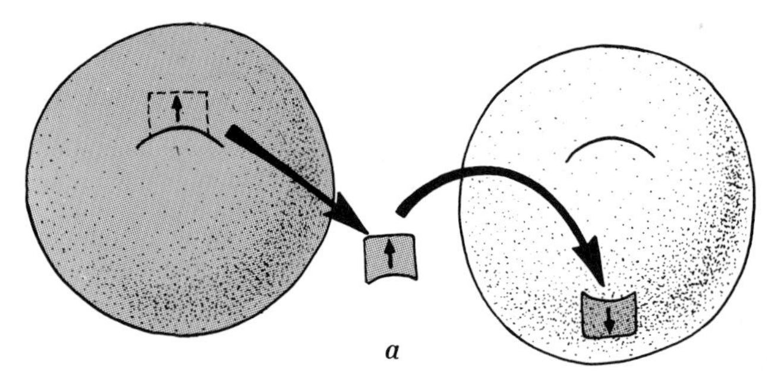

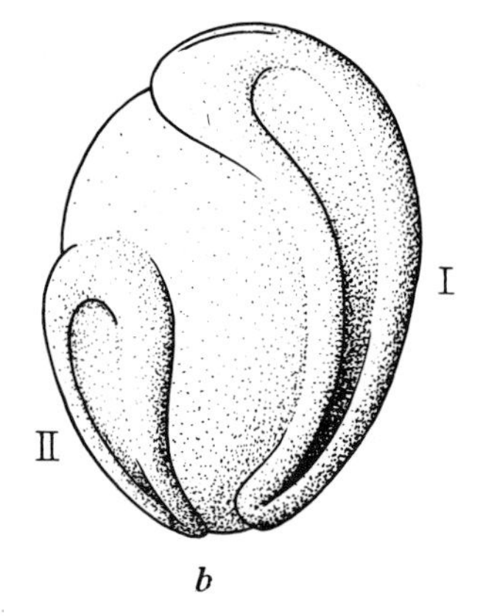

The fragment removed represents the notochord primordium. Grafted on another embryo, it develops normally and induces the neural and mesodermal structures of a second embryo.

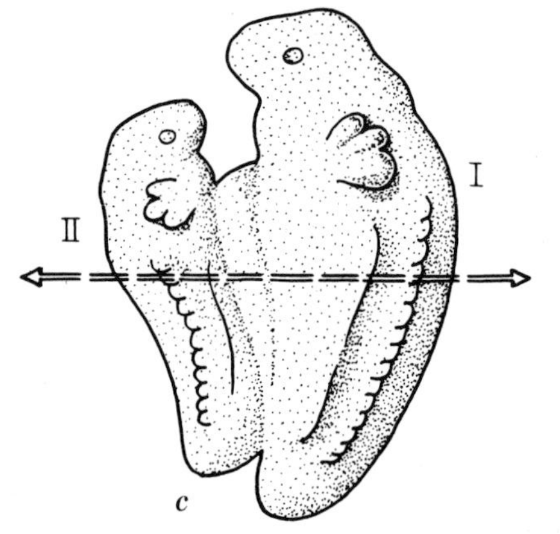

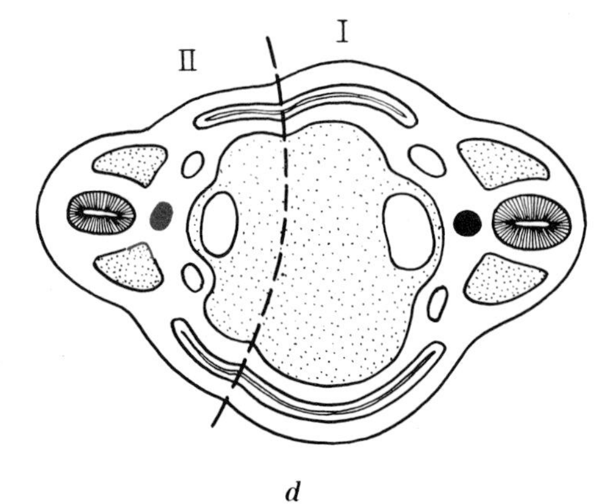

Fig. 1.

These experiments also clarified another major concept, that of competence. In order to react t the inductive effect, the reacting or competent tissue must already have attained a certain stage o development. However, in order for it to be qualified to react, its development must not already b fixed in the sense of specific differentiation. Thus, to be effective, induction must act at a specifi moment, neither too early nor too late.

Beginning with this primary induction, a series of secondary, then tertiary inductions are set i motion; the tissues first induced can in turn become inductors.

Secondary induction. — This can be shown by using as an example development of the vertebra column.

The vertebrae are formed from the sclerotome and include a vertebral body centered on the noto chord and a posterior arch surrounding the neural tube.

Ablation experiments on the notochord or on the neural tube provided information on the re spective inductive roles of these two structures in formation of the vertebrae (fig. 2).

. The notochord induces formation of the vertebral bodies from the sclerotome migrating from the somite.
The neural tube induces formation of the posterior arch from the surrounding mesoderm

Posterior vertebral arch
Neural tube
Notochord
Vertebral body
A
B
①

. Ablation of the notochord brings about anomalous development of the vertebral bodies which are reduced to a continuous, nonsegmented bony sheet. The posterior arches, however, form normally around the neural tube.

②

. Ablation of the neural tube hinders appearance of the posterior arches, while the vertebral bodies are formed normally around the notochord.

③

Fig. 2.

Nature of the inductor. — In numerous experiments, attempts have been made to demonstrate the organizer and its mode of action. The most informative are those using tissue cultures in the Grobstein apparatus. This method has been used to study inductor systems in many organ primordia. It has allowed demonstration of frequent inductive interactions between the mesenchymal and epithelial constituents of various primordia.

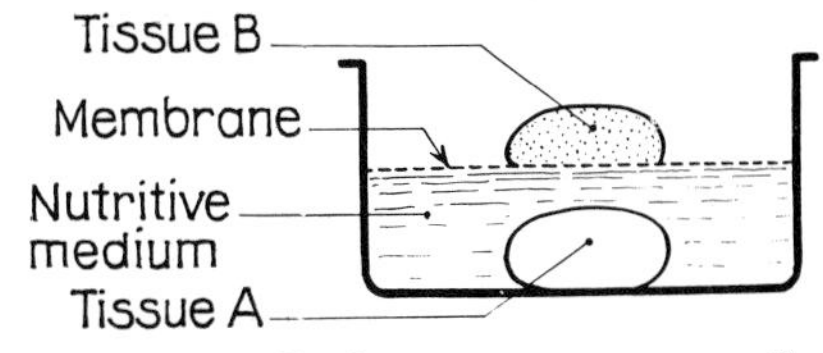

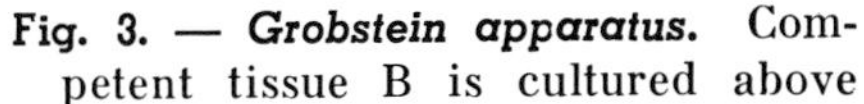

Fig. 3. — ***Grobstein apparatus.*** Competent tissue B is cultured above inductor tissue A, but is separated from it by a membrane impermeable to cellular elements, although it lets soluble substances pass easily. Differentiation of B proves that the inducing action is not exercised directly, but through the intermediary of one or more soluble substances which are at present still unknown.

The effect of induction is to provoke cellular differentiation. This is the morphological translation, at the cellular level, of synthesis of a new type of protein at the molecular level.

Thus, when the sclerotome begins to form the cartilaginous matrix of the vertebral premordium, this change in cellular morphology is preceded by synthesis in the mesenchymal cells of the constituent proteins of cartilage.

MECHANISM OF PROTEIN SYNTHESIS
(fig. 1)

Protein specificity is determined by the genetic material, that is, by the DNA (deoxyribonucleic acid) of the chromosomes, and more precisely, by the order of succession of the 4 bases which are part of this DNA = adenine, thymine, cytosine, and guanine *(arrow 1)*.

In the first phase, *transcription* occurrs : messenger RNA (ribonucleic acid) copies the segment of DNA concerned, respecting the base linkages *(arrow 2)*. Then this messenger RNA passes from the nucleus into the cytoplasm *(arrow 3)*.

The second phase is that of *translation* : the messenger RNA, carrier of the information, is fixed on the ribosome which is also formed largely of a specific RNA and which represents the site of protein synthesis *(arrow 4)*.

The ribosome receives, from a third type of RNA, transfer RNA, the materials necessary for synthesis, that is, the amino acids *(arrow 5)*.

Messenger RNA is thus decoded at the level of the ribosome, which it covers from one end to the other. Transfer RNA's are called progressively as their codes are uncovered, and deposit their amino acids in the indicated order. Thus, parallel with the passage of messenger RNA in the ribosome, a protein molecule is built up by linkage of amino acids *(arrow 6)*. Actually, the messenger RNA works simultaneously on several ribosomes (together forming a polysome), thus increasing the rate of synthesis.

In the genetic message, each amino acid is designated in a precise and constant way by a combination of 3 of the 4 fundamental bases (triplet code).

INITIATION AND REGULATION OF PROTEIN SYNTHESIS

Although the mechanism of protein synthesis is quite well known, there is still little information regarding its initiation in response to an inductive effect of extracellular origin.

Thus the general scheme of cellular differentiation remains to be completed.

)F DIFFERENTIATION

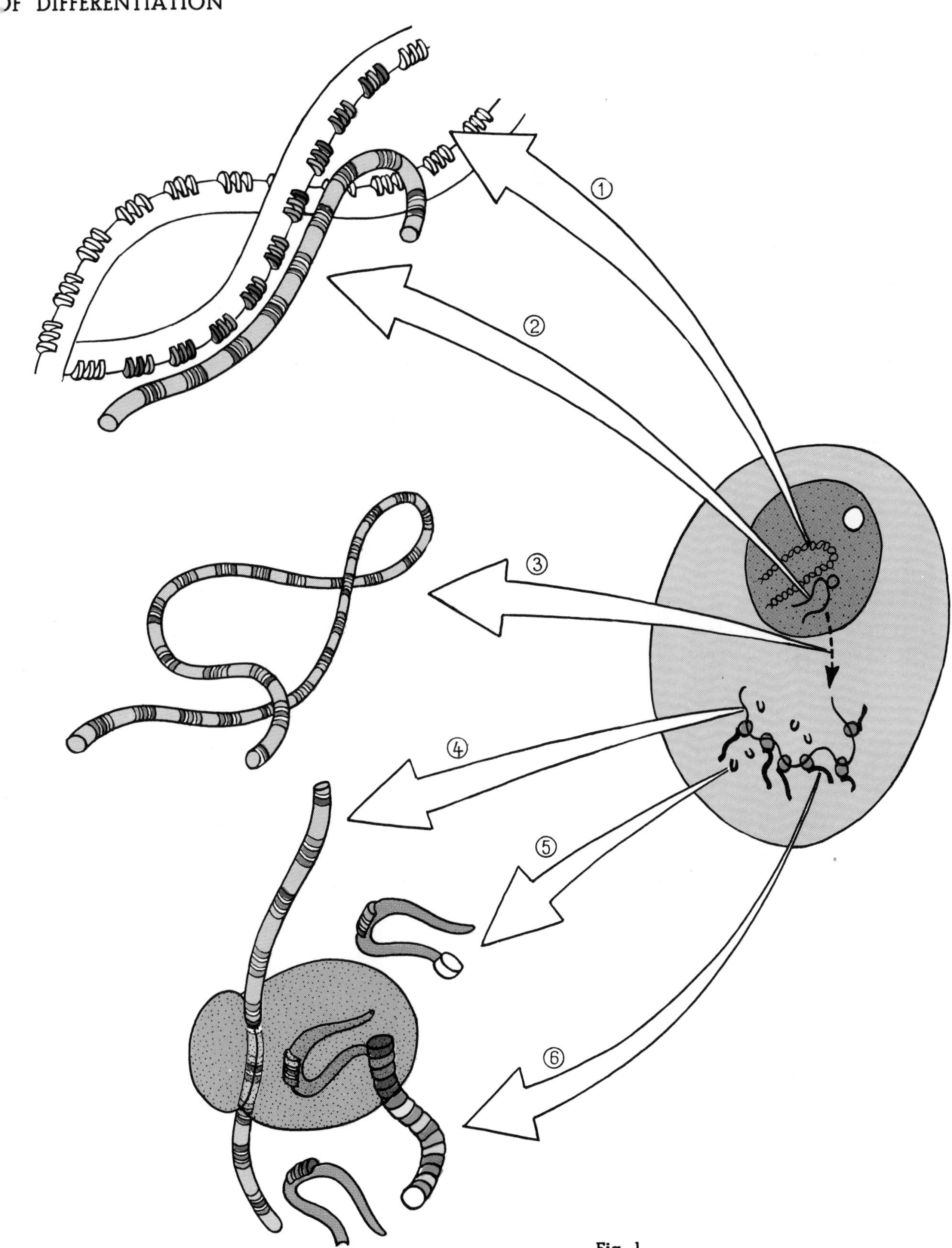

Fig. 1.

ANOMALIES

ETIOLOGIC FACTORS

Malformations arise from three mechanisms :—

— noxious influence of external factors during first phases of development;
— transmission of a genetic abnormality by the parents;
— chromosome aberration existing in one of the gametes or appearing during the first divisions

1. ***Noxious external factors.*** — This type of teratogenic agent can be studied experimentally. Their nature and mode of action will be considered later (p. 102).

2. ***Transmission of a genetic abnormality.*** — The anomaly is inscribed in the genetic code of one or both parents, and is transmitted according to the laws of heredity.

The fetus can be the carrier of a purely molecular abnormality : for example, *hemophilia* is the absence of a globulin necessary for coagulation.

Sometimes this molecular abnormality is translated on the cellular scale : for example, an *abnormal hemoglobin* can bring about a visible deformation of the red blood cells.

Finally, more rarely, certain metabolic anomalies bring about a morphologically discernable malformation : for example dwarfism and body deformities of *achondroplasia*.

3. ***Chromosome aberration.*** — We have seen earlier (p. 5) the mechanism which can cause an abnormality of chromosome constitution at the time of gametogenesis.

At fertilization, the zygote resulting from this gamete will have an abnormal karyotype (see fig. 1) which will be transmitted to all the cells of the embryo. The abnormality can involve an autosome or a sex chromosome.

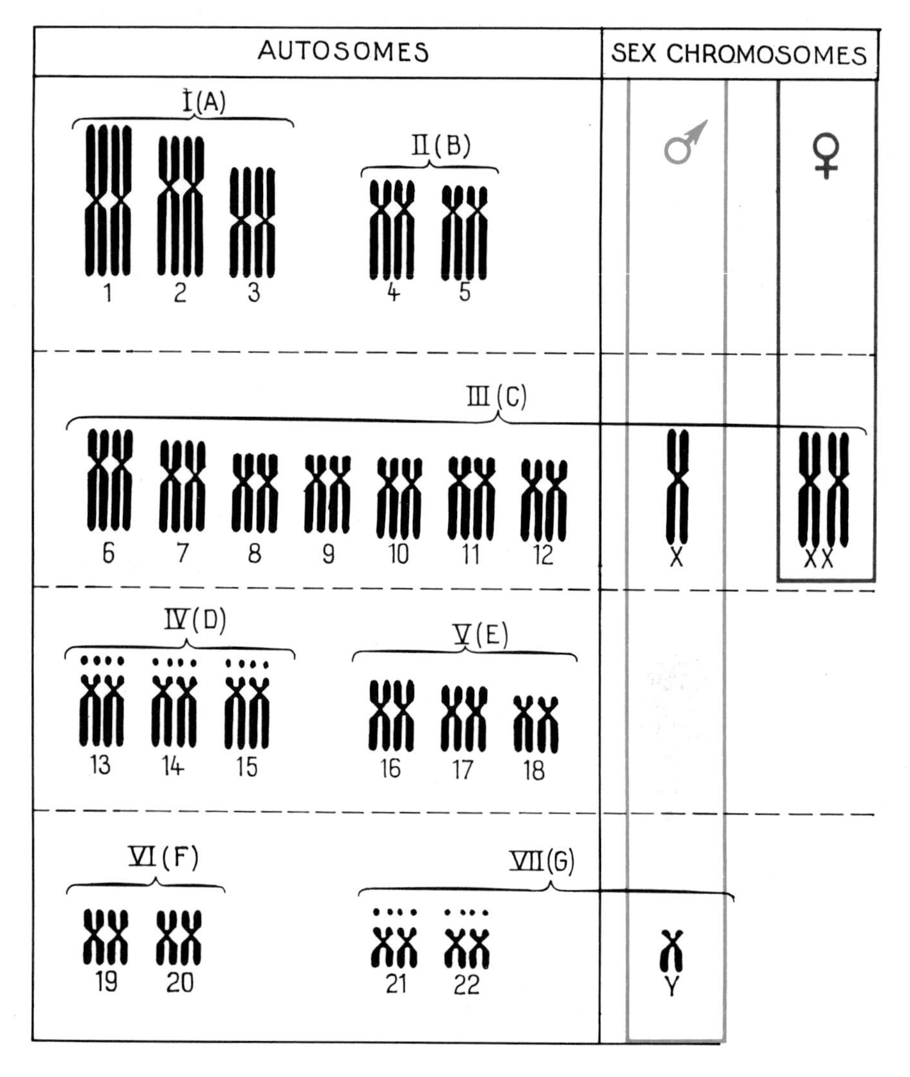

Fig. 1. — *Normal karyotype according to Denver convention, 1960.*

)F DEVELOPMENT

AUTOSOMAL ABERRATIONS

The most frequent autosomal aberration is ***mongolism*** (Down's yndrome), caused by trisomy 21 (chromosome pair 21 consists of 3 hromosomes instead of 2).

One gamete, here the ovum (as is most frequently the case) abnor- ıally carries 2 chromosomes 21 (fig. 2). The spermatozoön normally arries one. The zygote thus contains 3 chromosomes 21 and 47 chro- ıosomes in total (see also p. 5).

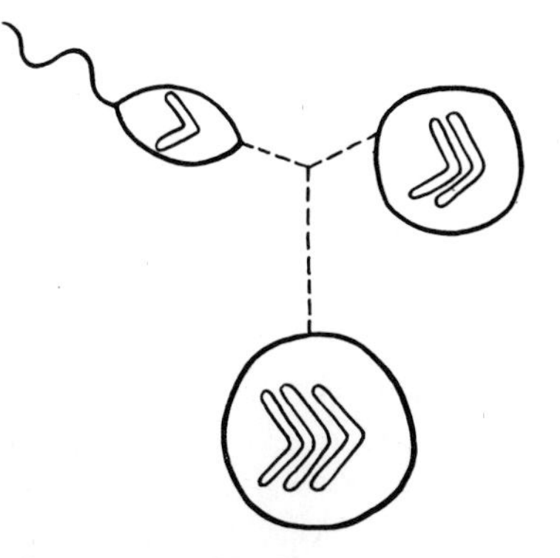

Fig. 2. — *Mechanism of trisomy 21.* Only the 21 pair is shown in this diagram.

SEX CHROMOSOME ABERRATIONS

Sex chromosome aberrations appear to be more frequent than autosomal abnormalities, but this ıay be because most autosomal anomalies lead to precocious death of the embryo.

Beginning with a chromosomal anomaly of the ovum, figure 3 explains the formation of the two rincipal sex chromosome aberrations : ***Turner's syndrome*** and ***Klinefelter's syndrome.***

In the first case, the ovum does not carry any sex chromosome. In the second case, it carries two.

These two syndromes are characterized by morphologic anomalies associated with gonadic dysgenesis (insufficient or abnormal development of the gonads).

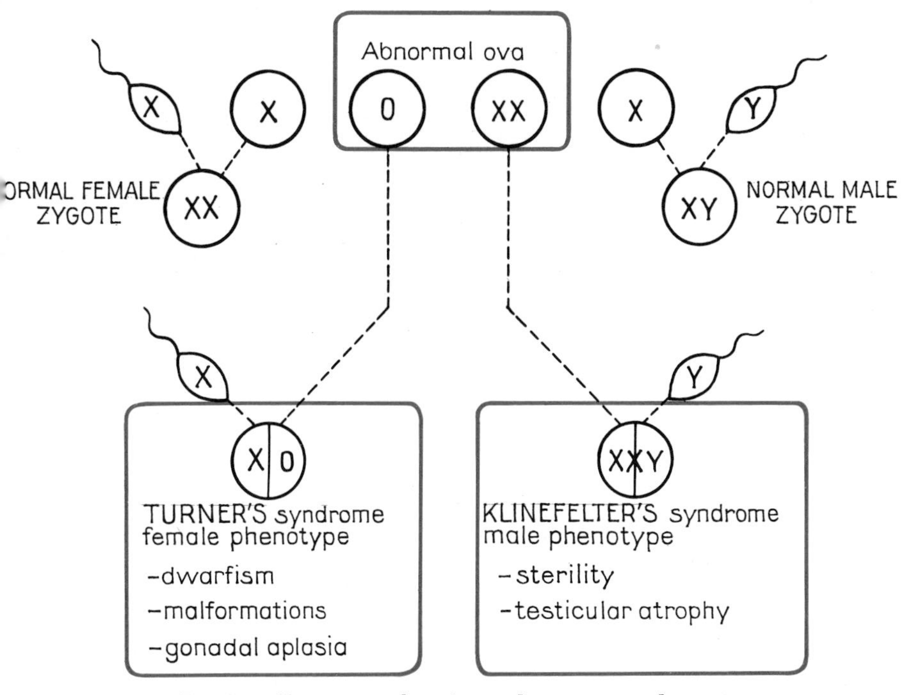

Fig. 3. — *Two examples of sex chromosome aberration.*

EXAMPLE

Malformations occur in 2 to 3 percent of human birth

I. — SINGLE MONSTERS

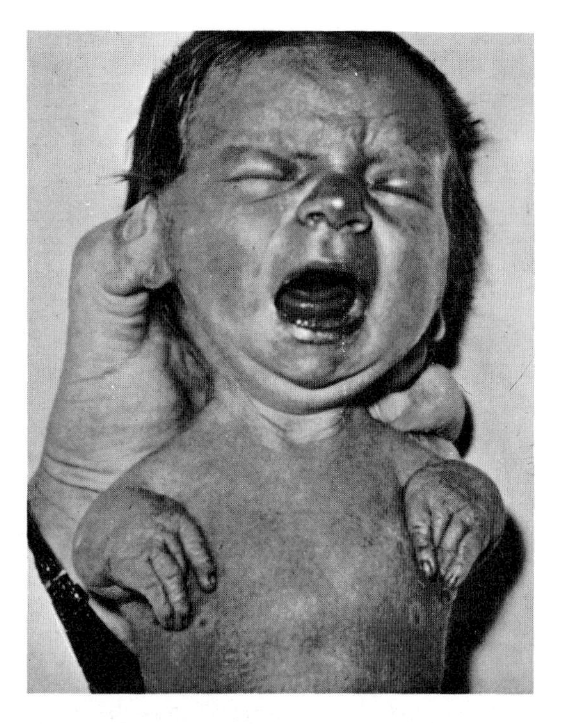

← Fig. 1. — *Phocomelia.*

Limb anomaly. Rare spontaneously (1/100,000 births). This is one of the typical lesions of thalidomide : 10 % of the women who took this drug during the critical period had babies with this characteristic anomaly.

(Photo kindly provided by Professors LEPAGE and SCHRAMM.)

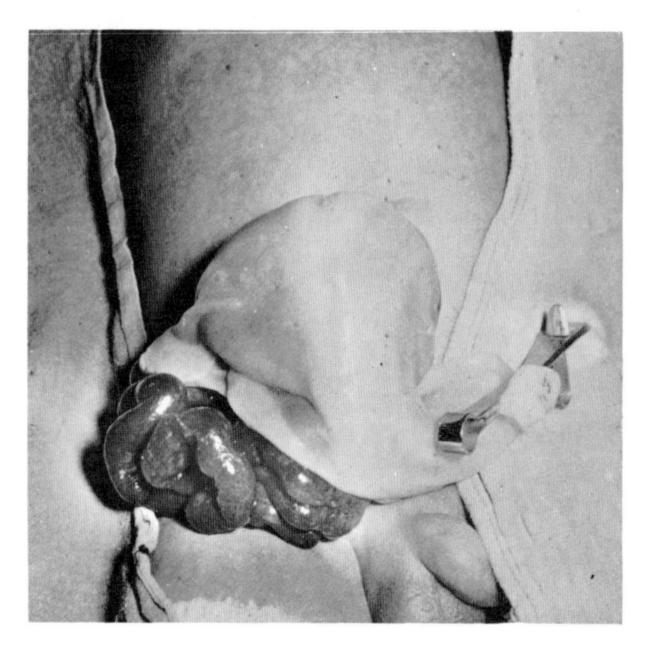

Fig. 2. — *Coelosomy.* Defect of closure of the abdominal wall : extraabdominal position of the viscera which are themselves normally développed.

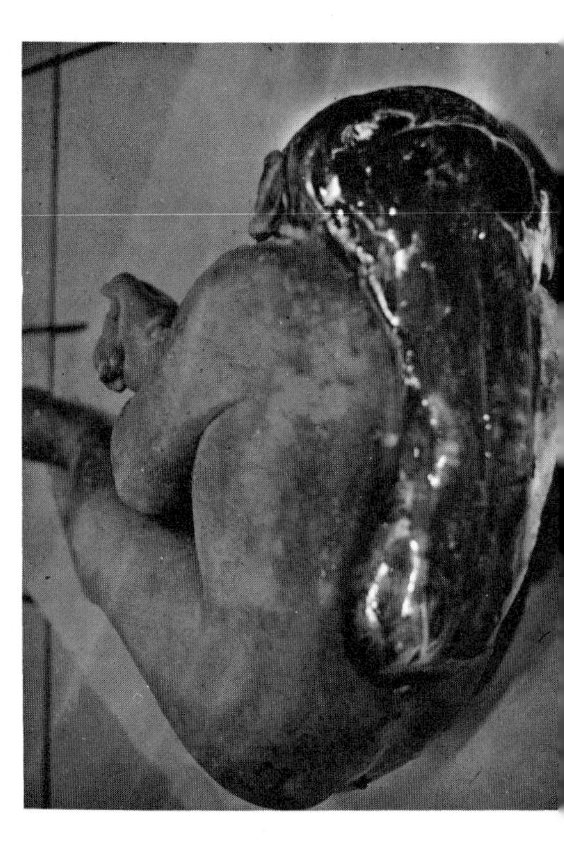

Fig. 3. — *Craniorachischisis.* Complete failure the neural tube to close.

Absence of cranial vault, absence of post rior arches of vertebrae. Angiomatous dege eration of nervous tissue.

OF HUMAN MALFORMATIONS

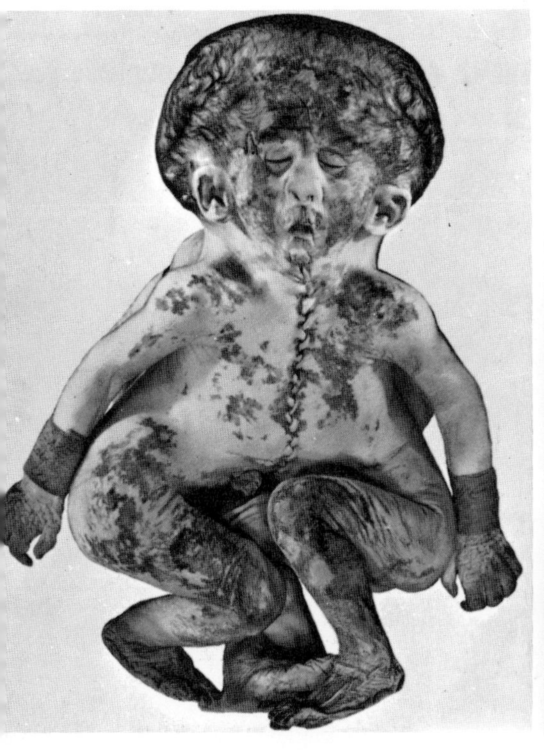

Fig. 4. — *Janus-type cephalothoracopage.*

II. — DOUBLE MONSTERS

Double-type malformations can be thought of as non-separated twins, the degree and type of fusion being very variable.

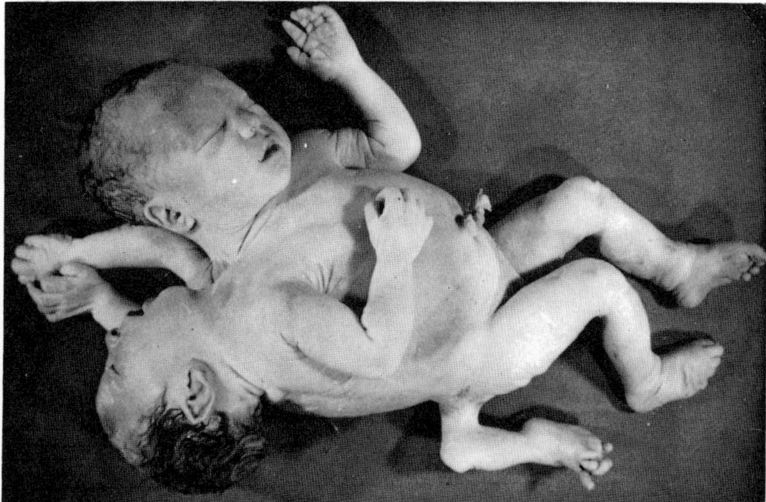

Fig. 5. — *Asymmetric thoracopage.*

Fig. 6 and 6 *a*. — *Acardiac.*

The acardiac is one of a pair of monozygotic twins. It has degenerated after a failure of vascularization. The structures already acquired regress and the process ends in the formation of a more or less amorphous mass, in which no general organization can be found.

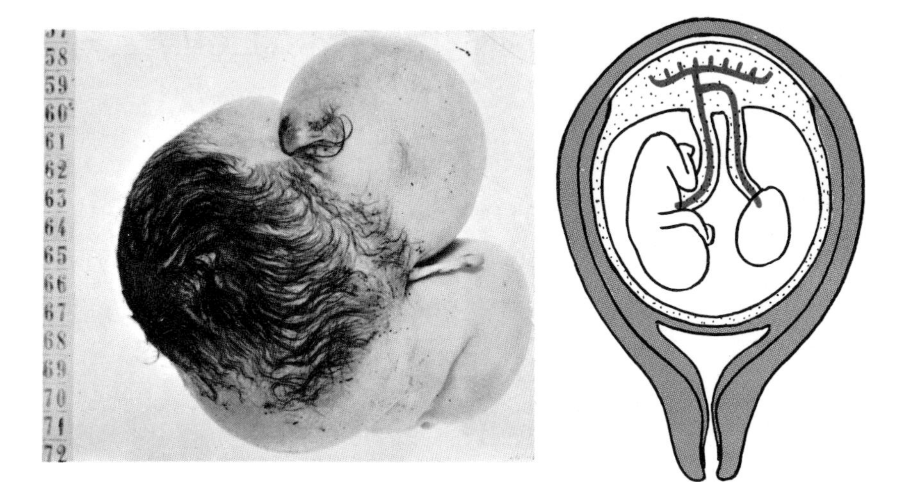

Fig. 6. Fig. 6 *a*.

EXPERIMENTAL

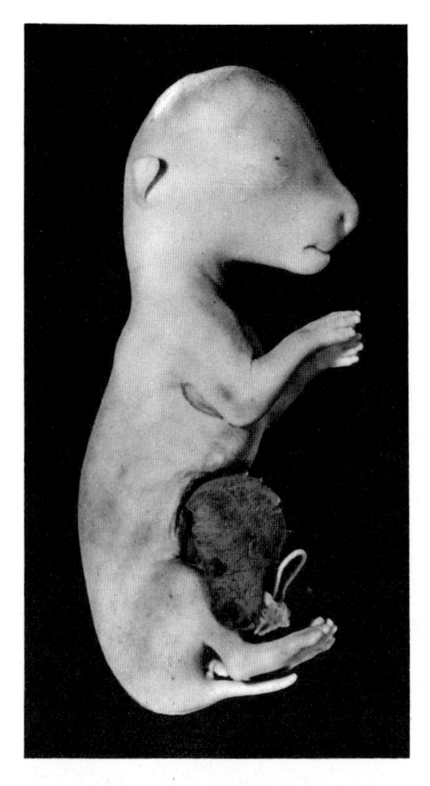

Fig. 1. — *Rabbit fetus.*
Coelosomy.

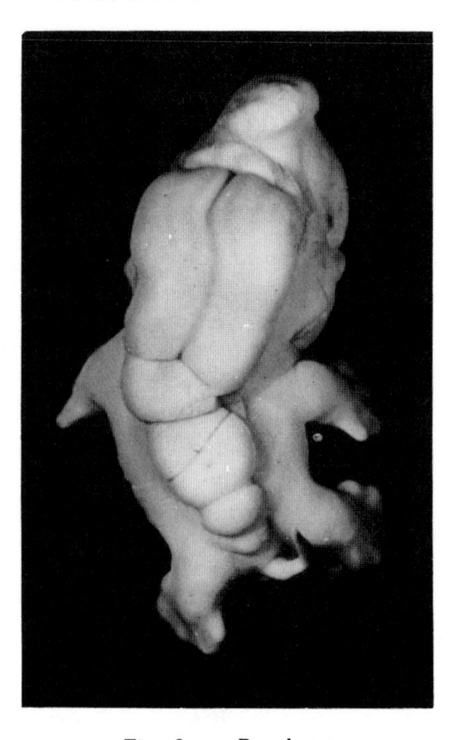

Fig. 2. — *Rat fetus.*
Craniorachischisis.

Despite its apparent protection, the mammalian embryo is very sensitive to the noxious influence of various external teratogenic agents. All the malformations observed in humans clinically have been reproduced in various mammals (fig. 1, 2, 3, and 4).

For example, the photographs opposite show experimental results comparable to the human malformations seen on the preceding pages.

CLASSIFICATION OF TERATOGENIC FACTORS

Teratogenic factors studied experimentally are classed in five groups :—

— physical factors : radiation, X-rays;
— chemical factors : antitumor, hypoglycemic, neuroleptic, etc.
— nutritional factors : vitamin imbalance : hyper- or hypovitaminoses; mineral deficiency or excess;
— hormonal factors : androgens, synthetic progesterones, cortisone;
— infectious factors : rickettsioses, toxoplasmosis, viruses (particularly rubella).

MODE OF ACTION OF TERATOGENIC FACTORS

The teratogenic effect depends essentially on the stage of intervention of the teratogenic agent (chronological factor) and on the genetic constitution (constitutional factor).

Several types of sensitivity can thus be outlined.

MALFORMATIONS

Time (stage) sensitivity. — 1. *Before implantation,* external agents, according to their intensity, provoke either completely reversible lesions, or definitive mortal lesions.

2. *After implantation,* and during the entire period of active morphogenesis, the principal teratogenic period occurs :—

— a primordium is most sensitive to teratogenic actions at the time of its appearance;

— the same substance can thus produce different malformations if it is administered at different stages of morphogenesis;

— when several primordia develop simultaneously (see pp. 88 and 89), the same substance can bring about multiple malformations.

Species sensitivity. — An agent teratogenic for one species may not be so for another, making interpretation of experimental results and their extrapolation to human treatment difficult.

Strain sensitivity. — In the same species, the percentage of malformations obtained with the same substance can vary considerably according to the strain and even the line.

Individual sensitivity. — In the same litter subjected to a teratogenic influence, certain individuals may be free of any malformation, and those which are malformed are not necessarily malformed to the same degree. It is probable that different metabolic peculiarities explain these variations between individuals and also between strains.

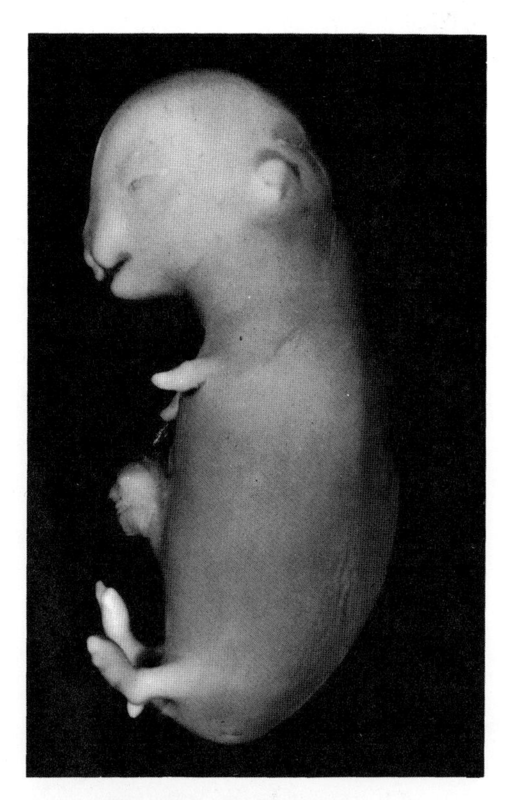

Fig. 3. — *Rabbit fetus.*
Phocomelia and harelip.

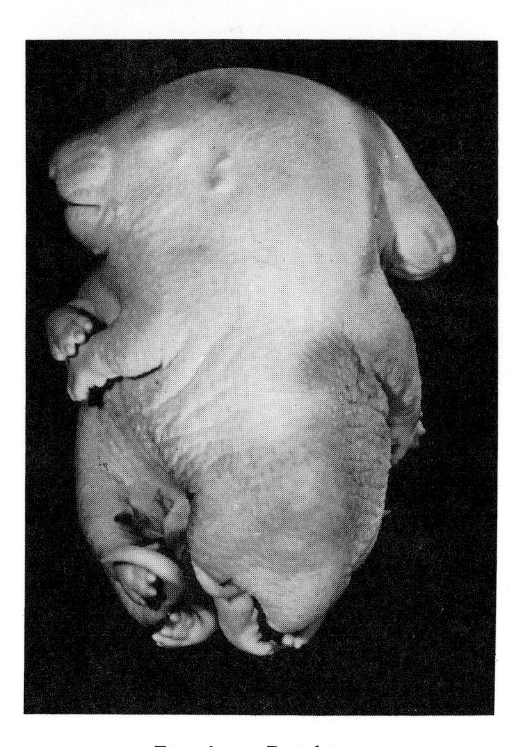

Fig. 4. — *Rat fetus.*
Cephalothoracopage Siamese twin.

FETAL-MATERNAL

The disease due to fetal-maternal incompatibility begins in the fetus and manifests itself after birth.

This is the most frequent fetal disease : about 1 in 150 births.

It results from incompatibility of the blood between mother and fetus. More exactly, in involves immunization of the mother against a blood group antigen carried by the fetus: the most frequent case is Rhesus (Rh) immunization.

Rh- red blood cell

TRANSFUSION

Rh+ red blood cell

anti-Rh antibodies

IMMUNIZATION

Fig. 1.

I. — MECHANISM OF Rh IMMUNIZATION

Rh is a blood group antigen carried by the red blood cells in 85 % of individuals (called Rh+). An Rh— subject can be immunized against the Rh+ antigen if Rh+ red blood cells are introduced into his body, for example by transfusion (fig. 1).

Immunization is then manifested by appearance of anti-Rh antibodies.

Immunization of an Rh— woman who has an Rh+ husband may occur during pregnancy. At this time fetal red blood cells may cross the placenta and enter the maternal circulation (fig. 3).

If the fetus is Rh+, Rh+ antigen is thus introduced into the maternal blood (fig. 2).

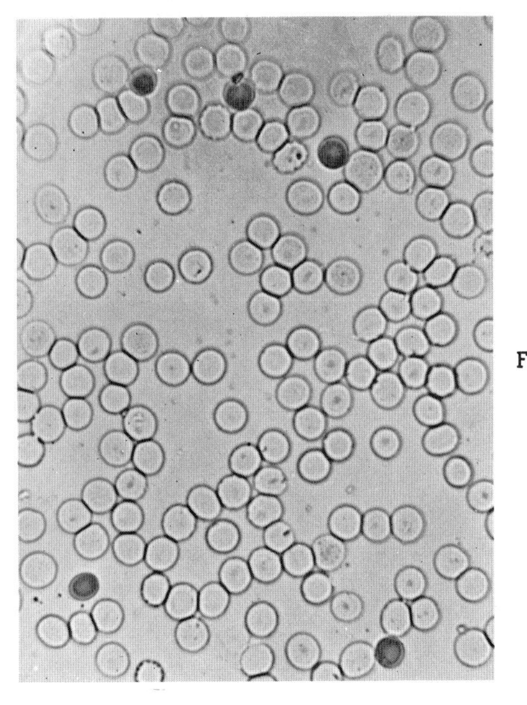

Fig. 2. — ***Maternal blood smear containing fetal red blood cells.*** This slide was treated with an acid solution which decolorized the maternal red blood cells, while the fetal red blood cells, more resistant to this treatment, remained normally colored. Five can be counted in this field.

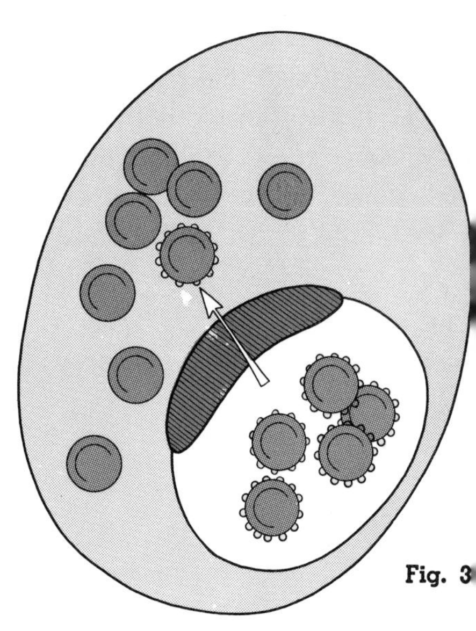

Fig. 3

NCOMPATIBILITY

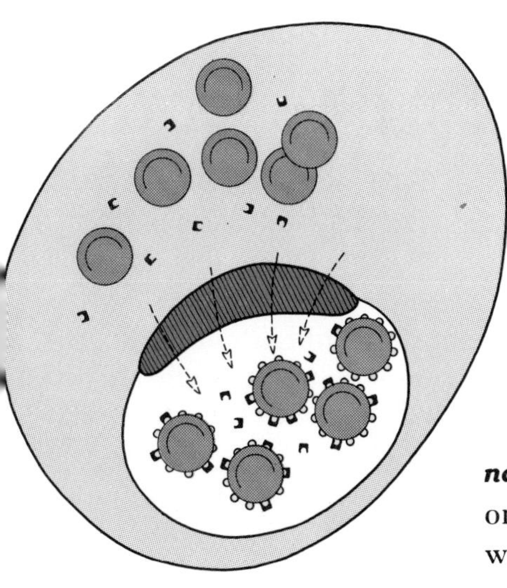

Fig. 4.

The mother is immunized against this Rh+ antigen and in her blood appear anti-Rh+ antibodies.

These, like almost all antibodies, easily cross the placental barrier and pass into the fetal blood (fig. 4).

In the fetal circulation, the maternal antibodies attach themselves to the red blood cells and cause their destruction in the spleen (fig. 5).

Thus ***hemolytic perinatal disease*** is caused; on the following page we will see its signs and symptoms.

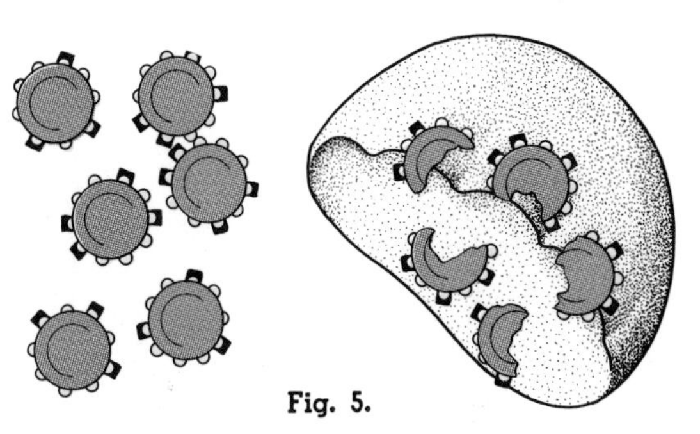

Fig. 5.

***

Let us summarize what can happen in couples where the mother is Rh— and the father Rh+ when an immunization occurs (which is far from the rule).

If the father is homozygous (fig. 6), all the children will be Rh+. The first pregnancy will be normal. Maternal immunization can appear from the second pregnancy on, and will become progressively greater with each pregnancy.

If the father is heterozygous (fig. 7), some children will be Rh—. These will be born completely normal, for the maternal antibody has no activity on their red blood cells. By contrast, the Rh+ children will be affected as described earlier.

Blood incompatibility affects only couples in which the mother is Rh— and the father is Rh+, and only 5 % of these.

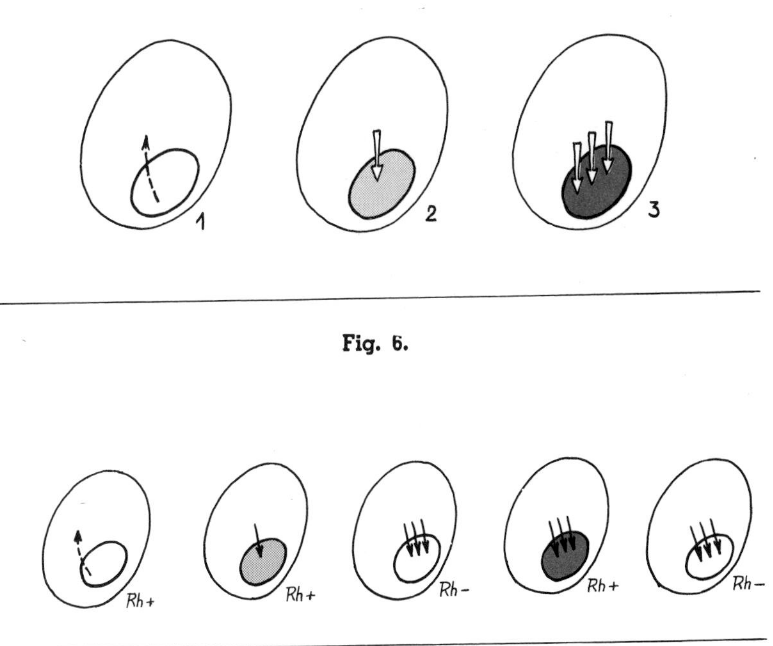

Fig. 6.

Fig. 7.

II. — CONSEQUENCE

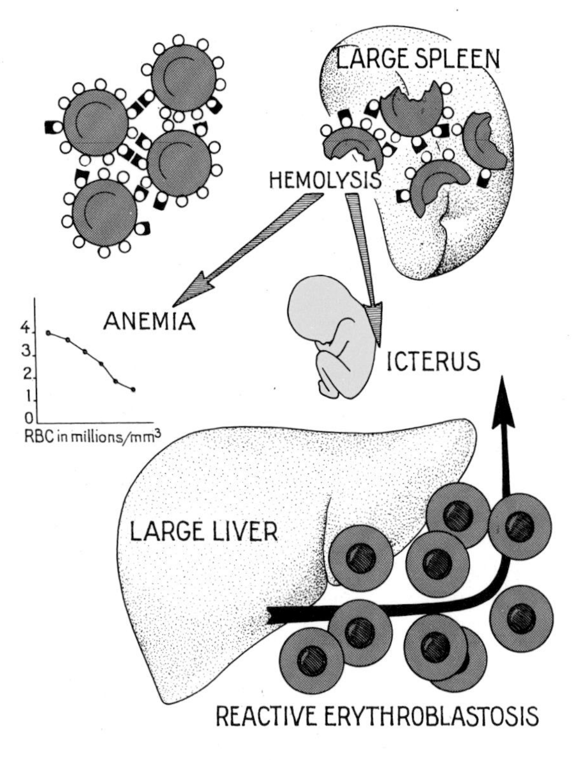

Fig. 1.

1. *Before birth.*

Maternal antibodies are fixed on the red blood cells and provoke their destruction in the spleen which hypertrophies.

From this results anemia and liberation of the hemoglobin pigment which is immediately denatured to bilirubin *(yellow)* (fig. 1).

This pigment crosses the placenta and is eliminated by the maternal organism.

The fetus reacts to the anemia by formation of new red blood cells. This erythropoiesis is carried out mainly in the liver (which also hypertrophies) (fig. 2).

A certain proportion of the red blood cells are young nucleated forms (erythroblasts) (fig. 3).

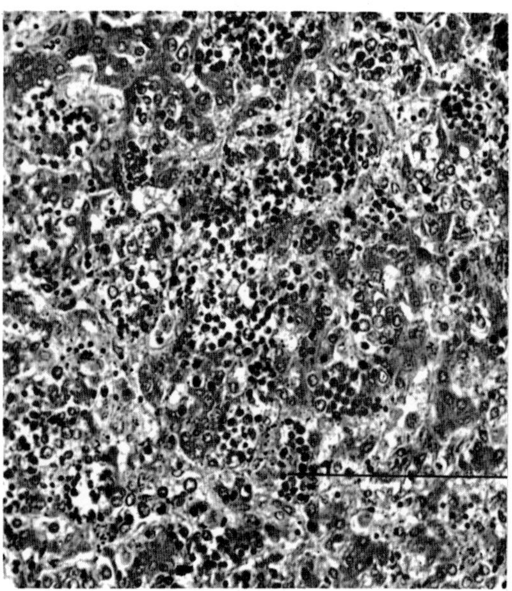

Fig. 2. — ***Liver section of a newborn, dead of hemolytic disease.*** The liver is still the site of intense erythroblastosis at birth.

Fig. 3 *(right)*.

Blood smear of a newborn with hemolytic disease. In comparison with normal blood *(lower insert)*, note :—

— the paucity of red blood cells;

— elevated proportion of erythroblasts.

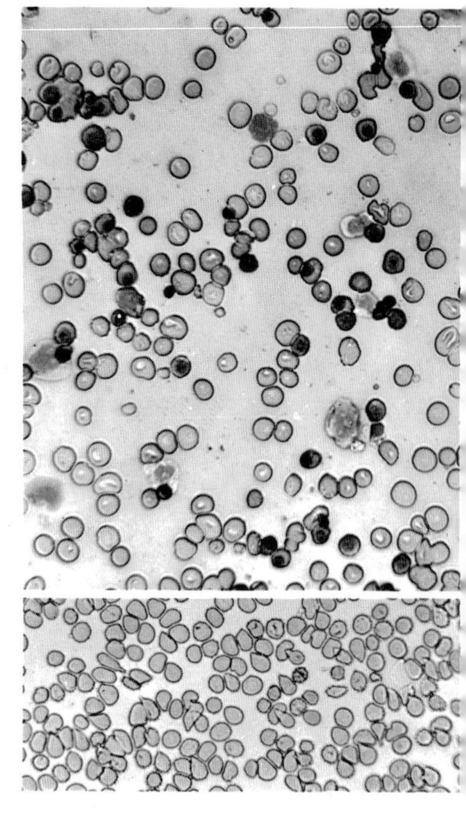

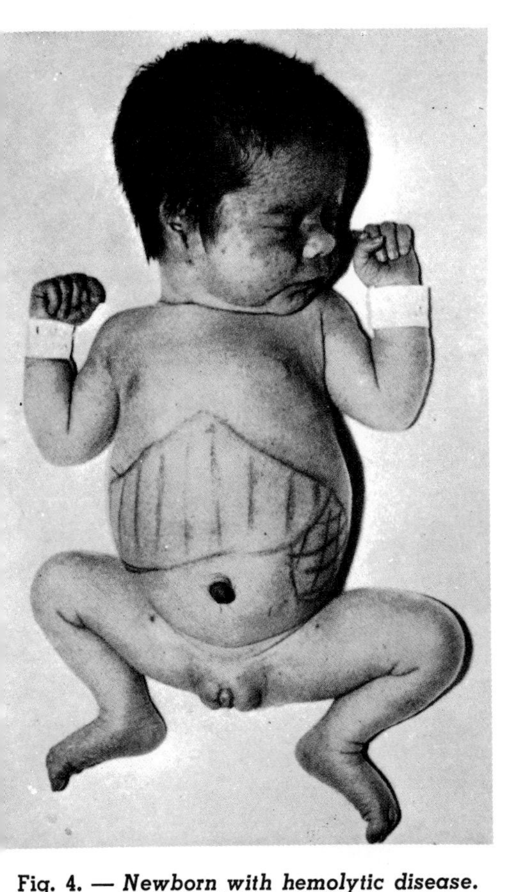

Fig. 4. — *Newborn with hemolytic disease.* The contours of the liver and the spleen, which are very hypertrophic, were traced with a dermographic pencil.

2. ***At birth.***

There is seen :—

— severe anemia with erythroblastosis;

— large liver and large spleen (photograph opposite);

— icterus which occurs very rapidly since the pigments are no longer being eliminated by the mother.

3. ***After birth.***

If the infant is not treated, he is threatened with death by anemia or by a specific complication, nuclear icterus, caused by the accumulation and toxic action of bilirubin in the gray nuclei of the brain (fig. 5).

Fig. 5.

4. In certain very serious forms, there is in addition to the preceeding signs, generalized fluid nfiltration : this is ***fetal-placental hydrops,*** which can produce fetal death in the last months of regnancy.

The placenta is almost as large as the infant (fig. 6). It can be more than twice the size of a ormal placenta (fig. 7).

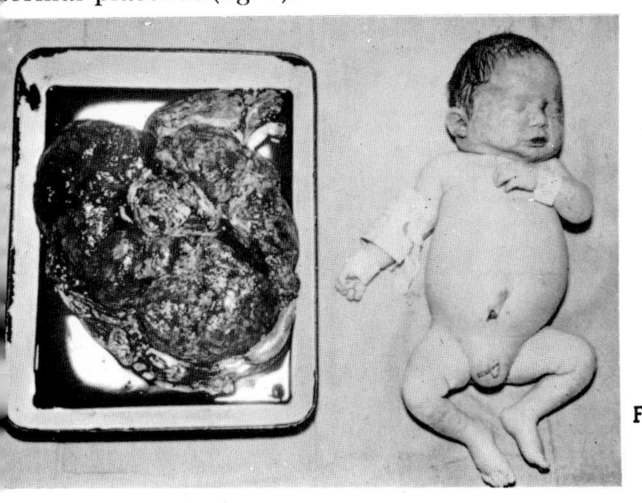

Fig. 6.

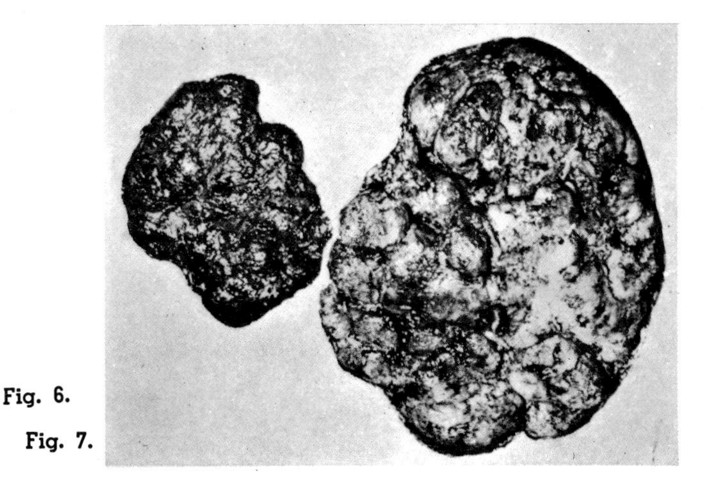

Fig. 7.

INDEX (*)

* Heavy type indicates main sections.

G

H

I

K

L

M

N

O

P

PRINTED IN FRANCE - Imprimerie Oberthur, Rennes - Dépôt légal n° 10.556, 2e trimestre 1975